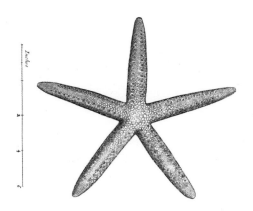

How Zoologists
Organize Things

How Zoologists Organize Things

The Art of Classification

DAVID BAINBRIDGE

FRANCES
LINCOLN

Brimming with creative inspiration, how-to
projects, and useful information to enrich your
everyday life, quarto.com is a favourite destination
for those pursuing their interests and passions.

First published in 2020 by Frances Lincoln,
an imprint of The Quarto Group.

The Old Brewery, 6 Blundell Street
London, N7 9BH,
United Kingdom
T (0)20 7700 6700 F (0)20 7700 8066
www.Quarto.com

A catalogue record for this book is available from the
British Library.

978-0-7112-5226-4

10 9 8 7 6 5 4 3 2

Typeset in Adobe Caslon and Bodoni

Senior Art Editor: Emma Clayton
Picture Researcher: Sara Ayad
Designer: Blok Graphic

Printed in Singapore

MIX
Paper from
responsible sources
FSC™ C007207

Page 2: John James Audubon (1785–1851), *The Birds of
America*, **1827–1838; Great blue heron.**

Front cover: Adolphe Millot (1857–1921), *Nouveau
Larousse illustré*, **1898, vol. 6, p.972 – fish. Robarts
Library, University of Toronto Library, via Archive.org**

Contents

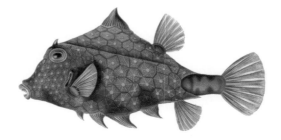

Introduction

Animals were among the first subjects drawn by the hand of man. Many early rock paintings are either representations of animals, or tracings of our thumby hands – a distinctively human attribute that sets our bodies apart from those of beasts, and *classifies* us as something different. We assume that our long-standing urge to make sense of the bewildering variety of animal life often had a practical basis – to differentiate the edible from the toxic, the ferocious from the tractable, for example – but there have always been artistic drives, too.

Long before Darwin, or Crick and Watson, our ancestors were obsessed with the visual similarities and differences between the creatures that inhabit the Earth alongside us. Early savants could sense there was an order, a scheme, that unified all life. And the classifications they formulated often tell us as much about humans' motives as they do about the animals they strove to organize.

Creatures were, indeed, organized in abundance. The human quest to classify living beings has left us with a rich artistic legacy in four great stages, in the West at least: the folklore and religiosity of the ancient and medieval worlds; the naturalist's cataloguing of the Enlightenment; the evolutionary trees and maps of the nineteenth century; and the modern, computer-hued classificatory labyrinth. Those four stages form the structure of this book.

We are told that zoological distinctions started early. On the fifth day of the Judaeo-Christian creation myth, God made the animals of the sea and air, but waited until the sixth to make the creatures of the Earth, man notably included. Even before Christianity took hold in Europe, zoological classification flourished around the Mediterranean – ancient Egyptians painted inventory-murals of edible animals and plants, while Aristotle formulated classificatory systems whose essence survives today. However, the surviving *visual* corpus of Western animal classification really begins during the High Middle Ages – some time in the twelfth century. Perhaps unsurprisingly, this early phase of bestiaries and encyclopedias reveals a world where animal diversity was forcibly reconciled with the Christian world view. For centuries, animals occupied immutable hierarchical levels of strictly gradated religious *scalae naturae* ('scales of nature'), while ferocious beasts – some real, some less so –

populated the terrifying margins of *mappae mundi* ('maps of the world'). Most of all, lavishly illuminated bestiaries initiated the compulsive cataloguing of animal forms that was to persist, in slowly evolving form, into modern modes of zoological organization.

The second phase in our story, the eighteenth century, brought a change of tone. Advances in art and science during the Renaissance, along with a yearning to return to an idealized conception of classical philosophy, had already led to new ways of seeing the world and its inhabitants. Now, animals began to be classified less according to the religious lessons they might teach us, and more by their objectively measurable similarities and

Jacob van Maerlant (c.1235–1291), *Der Naturen Bloeme,* **c.1350; Bird, with teeth.**

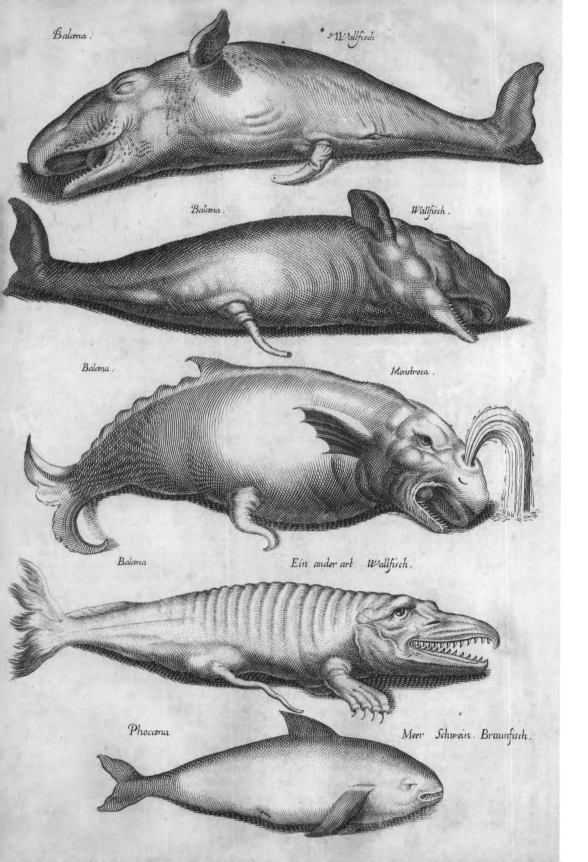

Balæna.   Wallfisch.

Balæna.   Wallfisch.

Balæna.   Monstrosa.

Balæna   Ein ander art Wallfisch.

Phocæna   Meer Schwein. Braunfisch.

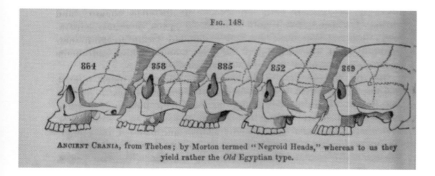

FIG. 148.

864   858   885   852   869

ANCIENT CRANIA, from Thebes; by Morton termed "Negroid Heads," whereas to us they yield rather the *Old* Egyptian type.

Josiah Nott (1804–1873) and others, *Types of Mankind*, 1854; Modern skulls – the fellahs of Lower Egypt.

differences. Close relations and distant relations, shared characteristics and distinguishing characteristics, possibly even ancestors and descendants – all pointed to underlying processes and hidden patterns of organization. Although seldom explicitly stated, it began to seem as if animals differ or are similar for reasons other than God's plan – maybe distinct animal types are actually *related* like members of a family; maybe they could even *change*. The incompleteness of these nascent ideas of natural history appears to have been a tantalizing challenge to the inquiring Enlightenment mind, and a flourishing of the naturalist/systematizer's art ensued, with creatures continually drawn, etched and painted into new organizational schemas.

Then, in the third phase, three new scientific insights were to drive the art of zoological classification in the nineteenth century. It was realized that animal species do indeed evolve – change and split – over time; Darwin and Wallace discovered the process of natural selection, which allows that evolution to take place; and the world was shown to be sufficiently ancient for evolution to have occurred, and indeed its rocks contain the neatly stacked fossil evidence of that process. Suddenly, evolution was real, it had a mechanism, and it had sufficient time, too. Yet it all seemed rather irreligious: all animals, including humans, had become linked by common ancestry, and geology, zoology and anthropology were united as well. As a result of these advances, nineteenth-century depictions of animal classification developed unmatched boldness and certainty. All nature could be elegantly and artistically summarized in a gnarled tree, a methodical table or an authoritative-looking map. Moreover, as Europeans diligently catalogued the fauna of distant continents, their native animals'

Opposite: Joannes Jonstonus (1603–1675), *Historiae Naturalis de Piscibus et Cetis Libri V*, 1650–1653; Whales.

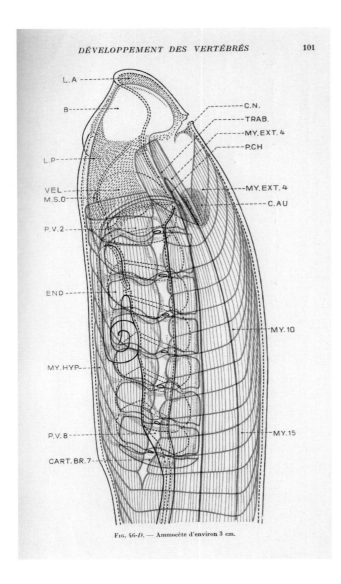

L.A

B

C.N.
TRAB.
MY.EXT.4
P.CH

L.P

VEL
M.S.O

MY.EXT.4
C.AU

P.V.2

END

MY.10

MY.HYP

P.V.8

MY.15

CART.BR.7

Fɪɢ. 46-D. — Ammocète d'environ 3 cm.

Pierre-Paul Grassé (1895–1985), *Traité de Zoologie, Tome XIII: Agnathes et Poissons,* **1958; Ammocoete.**

strangeness and variety confirmed the new theories being formulated 'back at home'.

The fourth phase of animal-organizing, the period since 1900, has been about much more than just accumulating more information. The deeper we peer into biological processes, the more meaningful animals' similarities and differences become. Species evolve and differentiate, but so do genes, chromosomes and genomes. Animals change, fossilize, adapt and interact in many ways – confusing ways that challenge the artist who depicts them. The neat evolutionary trees have become dense thickets, and the quest to untangle animal relationships has led to a range of strangely named disciplines – phylogenetics, taxonomy, chromosome mapping, phenetics, systematics, biostratigraphy, taphonomy, genomics. Compounding this complexity is the realization that animals interact in non-evolutionary ways as well – ecology, behaviour, symbiosis, parasitism, biomechanics, biophysics, environment and extinction. Yet now the richness of the scientific information is almost overwhelming, we can render it artistically beautiful in our attempts to tame it. As a result, the recent history of the art of animal classification has been the most diverse of all.

Again and again, we will see that illustrating animal variety leads people to do more than is strictly necessary. Sometimes the meticulousness of zoological cataloguing can seem pathologically obsessive, and the information it generates fills libraries with so many beautiful images that we rarely have time to look at most of them. Also, the artistic beauty of those images often exceeds what is required merely to report data or support a scientific theory. Again and again, there is descriptive and artistic overkill, as if a superabundance of striking visual representations may in itself somehow guide us towards deeper philosophical truths.

So the aim of this book is to tell the story of our artistic systematization of the beasts. Although its images and charts of the zoological world always parallel prevailing artistic trends and scientific discoveries, there are also strong conceptual threads that run throughout: animal life as a parable, a tree, an inventory to be catalogued, a network, a maze indeed, a *terra incognita*, and a mirror upon ourselves.

A clarification

The science of zoological organization has, unfortunately, acquired an array of overlapping jargon terms that can confuse not only the general reader, but many working biologists, too.

1 For most purposes, the terms 'classification', 'taxonomy' and 'systematics' may be assumed to mean the same thing – the identification of animal species and their allocation to a position in an organizational scheme. Scientists used to argue that these three words mean subtly different but overlapping things, but few worry about these distinctions now.

2 The word 'phylogenetics' refers to attempts to classify animals according to their common ancestors and genealogical relatedness. In other words, it assumes that animals

diversify by evolution. 'Phylogeny' means 'the origin of the race'.

3 The word 'phenetics' refers to the grouping of animals simply by their similarity or dissimilarity, with no implication they are related. Phenetic approaches were popular before evolution was generally accepted, but they are also used today when biologists do not think their data are adequate to discern true evolutionary relationships.

4 The phrase 'evolutionary biology' refers to the study of the mechanisms of the evolutionary process itself – the origin of life, the change of species over time, the splitting of species, and the external factors that influence these processes. It does not usually refer to the detailed process of actually classifying animals.

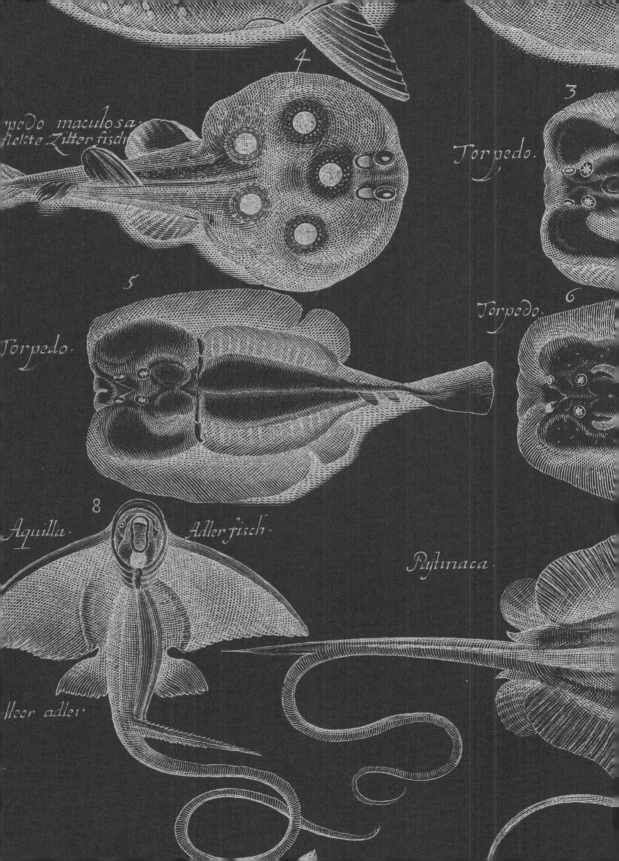

Torpedo maculosa
flekte Zilter fisch

Torpedo.

3

4

5

Torpedo.

Torpedo.

6

Aquilla.

Adler fisch.

8

Pastinaca.

Meer adler

Zitterfisch.

Angel fisch
Meer angel.

Aristotle, Bestiaries & Cynocephali

Joannes Jonstonus (1603–1675), *Historiae Naturalis de Quadrupedibus Libri*, 1657; Sunfish and rays.

An ABC of Early Classification (Antiquity–1700)

And out of the ground the Lord God formed every beast of the field, and every fowl of the air; and brought them unto Adam to see what he would call them: and whatsoever Adam called every living creature, that was the name thereof. And Adam gave names to all cattle, and to the fowl of the air, and to every beast of the field; but for Adam there was not found a helper fit for him.

Genesis, Chapter II, XIX–XX

The oldest Western traditions of naming and classifying animals arose from the Judaeo-Christian tradition, and zoological organization certainly has a prominent place in the Scriptures. Adam was created on the same day as the beasts of the Earth – a surprising nod to the modern concept of 'man as an animal' – and one of the very first tasks assigned to him was to name those beasts. Indeed, it

Anonymous, *Aberdeen Bestiary*, c.1200; God creates the animals.

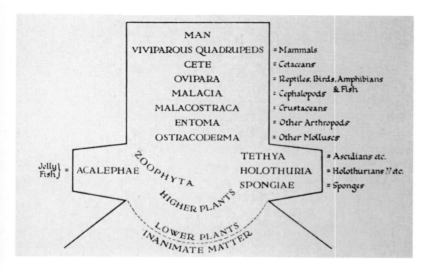

MAN	
VIVIPAROUS QUADRUPEDS	= Mammals
CETE	= Cetaceans
OVIPARA	= Reptiles, Birds, Amphibians & Fish
MALACIA	= Cephalopods
MALACOSTRACA	= Crustaceans
ENTOMA	= Other Arthropods
OSTRACODERMA	= Other Molluscs

Jelly Fish } = ACALEPHAE | TETHYA | = Ascidians etc.
HOLOTHURIA | = Holothurians ?? etc.
SPONGIAE | = Sponges

ZOOPHYTA

HIGHER PLANTS

LOWER PLANTS

INANIMATE MATTER

Charles Singer (1876–1960), *A Short History of Biology,* **1931; Aristotle's** *Scala Naturae.*

is notable that God only decided to create Eve because his non-human creations were not up to the task of accompanying and helping Adam.

This biblical trope of animal classification continues later in, and is perhaps partly explained by, detailed proscriptions against eating certain animals. Leviticus, Chapter XI contains what we would now call 'decision trees' relating to the eating of animals with cloven hooves, that chew the cud, or that have fins and scales. It is possible that these rules derived from earlier trial-and-error experiments with different foodstuffs, which led, presumably, to some disastrous microbiological or parasitic consequences, and these rules survive in modern Jewish customs. The reader cannot help feeling that many of these arcane injunctions must have had a practical life-or-death rationale, and indeed practical considerations have often driven our need to classify animals. After all we, by definition, are the descendants of people who *knew* which beasts were ferocious or toxic.

The other ancestral strand of Western zoological classification is the ancient Greeks, whose writings seem to spring from an enlightened inquisitiveness rather than a crude need to survive. Indeed, at first sight, their approach appears surprisingly modern. Much of early Greek animal biology is summarized in Aristotle's fourth-century BC *Natural Philosophy,* although it can sometimes be difficult to distinguish Aristotle's own discoveries from those of his often-uncredited sources.

Aristotle lived and wrote on Lesbos, one of the largest islands in the Aegean Sea, and his descriptions of the island's fauna, especially those that inhabited its warm, shallow lagoons, form the basis of much of the *Natural Philosophy.* Aristotle continually emphasized that his writings

were based on observation of nature itself and not a repetition of the
errors of his predecessors, and to some extent this is true. For each ζῷον
('zoön' means 'animal', hence 'zoology') he encountered, he analyzed
particular traits he thought could be used to determine its affinities with
other creatures. He realized that some features are common to all animals,
and also that colour, shape and size are unreliable classificatory criteria. As
a result, he recommended organizing animals according to the following
categories: their food, habitat and behaviour, how they breathe, whether
they metamorphose, whether they are social or solitary, nocturnal or
diurnal, tame or wild, offensive or defensive, whether they lay eggs or
bear live young, whether they are anchored to the seabed or swim, walk,
wriggle or fly free.

Yet Aristotle's thinking was not quite as modern as sometimes
credited. Although he produced the first known 'scientific' classification
of animals (see page 15), he was still bound by the wider metaphysical
structure into which he wished to fit the world. He was an early
proponent of the *scala naturae*, the ascending 'ladder of nature' by which
all things are ranked in a carefully gradated ladder of ascent from base
matter, through plants, animals and humans to the divine. This strict
hierarchical classification, although at odds with the sprawling animal
diversity Aristotle himself described, was to form the basis of many later
zoological classifications, and indeed philosophy in general until the
nineteenth century. It placed man reassuringly above the beasts, and set
him (for it was usually assumed to be a 'him') on a journey along a scale

of progress towards perfection, away from the formless and base towards the divine. However, Aristotle himself provided evidence that the *scala* was an imperfect concept – he often seems to consider humans to be 'just another animal', for example – yet it proved to be one of his most tenacious ideas.

Written by an unknown hand, probably in the second century AD, in Alexandria, the *Physiologus* is the next waypoint on the journey to modern animal classification. In many ways, it set the tone for over a thousand years. It is infused with the philosophy of the new Christian religion, and the forty animals it describes are presented less according to their zoological attributes, and more for what they can tell us as religious symbols. Each creature acquires a role in the Christian story, drawing on the earlier Greek traditions of animal parables, to illustrate particular theological principles. In the *Physiologus*, animals serve merely as illustrative elements of the word of God – zoology had become subservient to theology, and for centuries the scientific coherence of the former discipline was to suffer as a result.

Anonymous, *Hereford Mappa Mundi*, c.1300; *Cynocephali*.

A strange detour in these zoological traditions came in the early seventh century with the *Etymologiae* of Isidore of Seville. Isidore's book was an enormous undertaking, ostensibly an early attempt to create an encyclopedia of all contemporary knowledge. Yet it was skewed by the author's opinions about which ancient ideas were deemed deserving of perpetuation. Also, as the book's name suggests, Isidore believed that discovering the origin of particular words was the key to understanding the ideas they denote. So, for example, he saw the concept of being 'elephantine' as directly philosophically equivalent to the very nature of the large grey pachyderm – whereas the modern reader would simply assume that concepts can be used to name animals (a sloth, for example), and animals can give their names to concepts (as in a 'dogfight'). Indeed, in the long story of animal classification, the *Etymlogiae* is perhaps the instance when the human urge to organize animals most outstripped our actual

Conrad Gessner (1516– 1565), *Historiae Animalium*, 1551–1558; *Rana perfecta*.

understanding of those animals. However, the tome was endlessly reproduced throughout medieval Europe and the Islamic world, and became immensely influential.

These, then, were the forerunners of animal classification in the Late Middle Ages. Aristotle's metaphysics, Isidore's obsession with names, and Christianity's inward-looking self-justification led directly to the first and one of the most artistically spectacular bodies of zoological organization: the bestiary.

Medieval bestiaries flourished, especially in France, England and Scotland in the twelfth and thirteenth centuries, and while their level of artistic achievement varies, they were surprisingly consistent in their structure and focus. Indeed, so much did their compilers draw on previous and contemporary works for their inspiration, that the evolution of medieval bestiaries can itself be arranged into an evolutionary genealogy, with works fitting neatly into 'family' lineages of relatedness and descent. The mainly north African animals of the *Physiologus* were supplemented with north European creatures, as well as mythical beasts, to yield a rich menagerie from which theological lessons could be extracted. Spectacular illustrations presumably made these parables accessible to the illiterate masses, and it is not known whether their audience worried about which of the animals depicted actually existed. In bestiaries animals have a meaning beyond their actual physical nature, so their existence or otherwise is less important than what they can tell us about God. Some symbolism was straightforward – a fox traps birds as the devil ensnares sinners; a panther mauls a dragon as Jesus attacks Satan. However, some animals, especially

Ulisse Aldrovandi (1522–1605), *Serpentum, et Draconum Historiæ*, 1640; Snakes and dragons.

those with which medieval readers were more familiar, could be complex characters – a goat might be a sinner swallowed by hell in one context, and then the all-seeing sage Christ in another.

Another visual format that became popular in the Middle Ages was the *mappa mundi*, in which stylized geographies of the entire world (i.e. God's creation) were summarized in giant cartographic images. In most, the three known continents, Asia, Africa and Europe, are arranged around a centrally placed Jerusalem, a reassurance of God's place at the centre of the world. Around their periphery, however, lie strange lands, inhabited by exotic and horrific creatures – often distortions of real animals formulated to scare the faithful, or even devilish animal/human chimaeric hybrids. Some of the most dramatic examples of this zoological otherness appear towards the edges of the *Hereford mappa mundi* (see page 26) where dog-headed people or cynocephali, perhaps based on accounts of real-life baboons, can be seen frolicking.

During the Renaissance, thinking began to change. The religious certainty of medieval cataloguing started to give way to more objective attempts at classification. The sixteenth century saw the publication of works such as the Swiss philosopher Conrad Gessner's meticulous *Historiae Animalium*. Fifty years later came Ulisse Aldrovandi, with his vast collection of zoological curiosities in Bologna, and his *De Piscibus* and *Ornithologiae* among many other books. The sheer variety of animal life was rendering old ideas of a neat *scala naturae* untenable. The 'steppiness' of the *scala*'s ladder of creation was starting to look less clear-cut. Indeed, its neat linear progression from base matter to godhead now seemed an oversimplification. One day this stepwise ascent of animal creation would be transformed into a branching tree, and other more alien forms never conceived by the medieval artist.

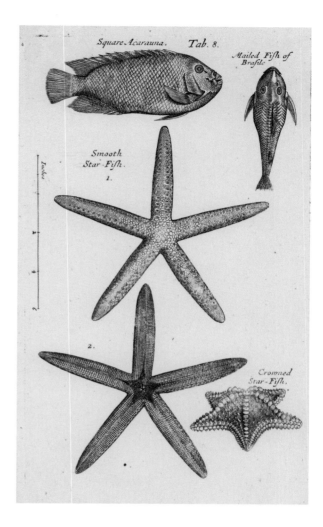

Nehemiah Grew (1641–1712), *Musaeum Regalis Societatis*, 1681; Fish and starfish.

Anonymous, *Aberdeen Bestiary*, c.1200; God creates the birds and fishes (above); Adam names the animals (opposite).

Although not the oldest medieval bestiary, the *Aberdeen Bestiary* is perhaps the most visually spectacular. Sometimes stirring, sometimes humorous, its illustrations were created to instruct readers in their quest to become closer to God. The creation of animals, and especially Adam's naming of the animals, are strangely prominent in the Genesis narrative (see page 14) – a foretaste of later thinkers' obsession with animal classification.

Anonymous, *Northumberland Bestiary*, c.1250–1260; Hedgehogs and bees (above and below); Sea monsters (opposite).

Bestiaries were a strange mix of the everyday, the exotic and the mythical – and the symbolism of some creatures may seem surprising to the modern reader. The *Northumberland Bestiary*, for example, depicts the harmless and charming hedgehog as an evil thief, using its spines to roll in fruit and steal as much of it as possible.

Noah's Ark

Medieval motifs of life, death and classification

T he origins of the Mesopotamian flood myths are unknown, but extremely ancient – possibly more than 5,000 years old. Although a global flood is impossible, these stories may derive from localized severe flooding of the Tigris and Euphrates rivers, or may even represent memories of great thaws following the last ice age – the modern Persian Gulf, for example, had once been inhabited dry land.

The flood story recounted in the Bible's book of Genesis is just one of several accounts, and is surprisingly brief, essentially comprising several repetitions of its first few lines:

Frère Laurent d'Orléans,
Somme le Roi, **c.1295;**
Noah's Ark.

And the Lord said unto Noah, Come thou and all thy house into the ark; for thee have I seen righteous before me in this generation. Of every clean beast thou shalt take to thee by sevens, the male and his female: and of beasts that are not clean by two, the male and his female. Of fowls also of the air by sevens, the male and the female; to keep seed alive upon the face of all the earth. (Chapter VII, I–IV)

However, the titanic scale of Noah's scanty story, and the comforting way a godly man categorized beasts and fowl, clean and unclean, male and female – and thus saved all living things – provided an irresistible trope for creators of medieval illuminated manuscripts.

A superb example is found in *Somme le Roi*, a guide to theological and moral rectitude compiled for King Philip III

of France by the Dominican monk Frère Laurent (see opposite page). Clearly, an ark was not an easy thing to draw, and each animal species ends up residing awkwardly in its own schematized monastic cell. It is notable that many medieval arks look more like modern scientific tables than anything that might effectively breast the waves of the deluge.

Above: Beatus of Liébana (c.730–785), *Commentary on the Apocalypse,* **twelfth century; Noah's Ark.**

Anonymous, *Hereford Mappa Mundi,* **c.1300.**

Of all medieval *mappae*, the one at Hereford is perhaps the most spectacular – 24 square feet of lushly inscribed and illuminated vellum. Like many medieval maps, the three known continents are arranged in a circle with Jerusalem at the centre, and exotic and mythical beasts and beings scattered around the periphery (see page 19). This image is of the 'two to three o'clock' section of the map, containing a heavily distorted Egypt and east Africa, inhabited by a mandrake, a salamander, a phoenix, a unicorn, a rhinoceros and a yale (an antelope with swivelling horns). There is even a golden calf for the ancient Israelites' misguided worship.

Bartholomaeus Anglicus (1203–1272), *De Proprietatibus Rerum*, **posthumously illustrated 1403; Lion, deer, unicorn, horse.**

Little is known of 'Bartholemew the Englishman' other than that he worked in Paris and Magdeburg, and wrote an early encyclopedia.

Ramon Llull (c.1230–c.1315), *Ars Magna,* **c.1305; Ladder of Ascent and Descent of the Mind.**

The *scala naturae*, or 'scale of nature' (see page 16), of the Majorcan philosopher Ramon Llull has been perhaps the most influential depiction of a stratified hierarchy of inanimate, animate and divine phenomena, setting the theme for many later biological 'great chains of being'. The implied sense of progress towards perfection was not to be expunged from biology until well into the twentieth century. The steps in Llull's philosophical staircase are rocks (*lapis*), fire (*flama*), plants (*planta*), beasts (*brutum*), man (*homo*), sky (*caelum*), angels (*angelus*) and God (*Deus*).

Didacus Valadés (1533–1582), *Rhetorica Christiana,* **1579; Great Chain of Being.**

Distant in time, but not theme, Didacus Valadés' scale of nature is more densely populated than that of Ramon Llull (shown opposite), but it is equally unyielding in its ordering of creation. Born in Mexico, Valadés was a Fransiscan monk whose theological treatise *Rhetorica Christiana* was influenced by his hierarchical view of the status and abilities of the unfortunate native inhabitants of Spain's New World dominions.

Jacob van Maerlant (c.1235–1291), *Der Naturen Bloeme,* **published with illustrations 1350; Various animals.**

Ultimately derived from the *Physiologus* (see page 17), thirteenth-century Dutch poet van Maelant's *Der Naturen Bloeme* ('The Flower of Nature') is ostensibly a description of the entire natural world, but focuses heavily on animals. The origins of this illustrated version, created after his death, are uncertain. Drawing on the bestiary tradition, the volume is profusely illustrated with creatures familiar, strange and mythical. March hares rub furry shoulders with the archetypal lion and lamb, walking fish, Bactrian camels, eagles and funnel-nosed elephants.

Hartmann Schedel (1440–1514), *Nuremberg Chronicle*, **1493;
Creation of the fishes and fowl (above); Creation of man and the
animals (opposite).**

A history of the world in seven ages (the seventh being the Last Judgement), the
Nuremberg Chronicle is one of the earliest printed books to combine images and
text. Its 'first age' contains unusually charming depictions of that primary biblical
classification of animals, the fish and birds created on the fifth day and humans
and land animals on the sixth.

Anonymous, *Miroir de l'Humaine Salvation*, c.1320 (this version c.1500); Noah's Ark, Jonah and the Whale, and other stories.

Many illuminated versions of the 'Mirror of Human Salvation' survive, recounting famous stories from the Old Testament, which are then claimed to prefigure events during and after the life of Christ. Over the centuries, the text was sometimes modified, and a wide variety of illustrative styles was employed. Once again, diverse animal types were to the fore as theological symbols or devices.

**Illustrations attributed to Robinet Testard
(1470–1531),** *Le Secret de l'Histoire Naturelle
Contenant les Merveilles et Choses
Mémorables du Monde,* c.1500.

Its different versions hard to attribute or date,
'The Secret of Natural History Containing the
Wonders and Memorable Things of the World' is
an early encyclopedia and gazetteer of the world,
often richly illustrated with a variety of creatures,
both real and apocryphal, in wild and distant
lands. Many of the tales are based on parables
dating back to the ancient Greeks.

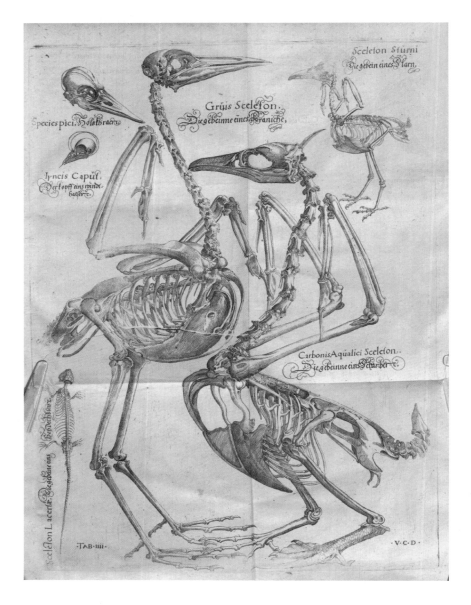

Volcher Coiter (1534–1576), *Lectiones Gabrielis Fallopii,* **1575;
Bird skeletons (above);** *De differentiis avium* **(opposite).**

Born in Groningen, Volcher Coiter's work represents an important
step towards modern methods of zoological organization. He
conducted detailed analyses of animal structure and himself produced
meticulous engravings, especially of skeletons. He also created detailed
animal classifications, repeatedly dividing and subdividing animal
groups according to their anatomical features – birds, in the example
illustrated here.

Volatu, aliæ

Volatiles aliæ, grandibus atq; ualidis sunt, utpote

Quæ ungues habent aduncos, & carne uescuntur, uolaces eris tota uita esse necesse est, quod ob unguium curuaturam ad currendum, & ut insistant (axis ineptæ sint. Vncunguium corpora sunt xigua exceptis alis : quoniam alimenti copia in alis & arma præsidio uniq; absumitur.

Quæ pernicitate sese tuentur.

Quæ loca mutare solent, ut sunt ardeæ, ciconiæ, grues, & idenua aliæ.

Non uolatiles, sed graues, quibus uita terrena, hæ autem

Fruge uiuunt, ut sunt gallinacea & omnia, perdices, cognices, & similes.

Nant, ut sunt mergi.

Apud aquas morantur.

Volandi modo, aliæ enim

Pedibus & cruribus exporrectis uolant, utpote longis cruribus præditæ, quæ cruribus loco gubernacu & caudæ utuntur, tales sunt grues, ciconiæ, ardeæ & similes alites.

Cruribus & pedibus ad uentrem contractis uolant, sic autem

Nullis partibus impedimento sunt pes.

Pedes uncunguibus ad raptum expeditiores.

Collo sunt longo. Cui aui collum est

Crassum simul & longum, exporrecto uolat collo.

Tenue & longum, curuato fertur collo. Cum enim tenue re longitudine quatitur, facilè frangi posset.

Collum habent breue, hæ expedito collo uolant.

Locis, in quibus uictitant, & maiori uitæ parte degunt. Quædam uiuunt in

Aquis & natant, hæ breuia habent crura, & palipedes, siue Plinio solidipedes sunt.

Paludosis & palustribus locis uersantur, quæ ad consequuta sint crura & colla longa, & pedes minimè membraneos. Nam ad nandum non sunt nata, degunt in solo præmoli & lubrico. Accommodauit enim natura instrumenta ad officia. Hinc ijs longa crura, digiti productiores & longè a se mutuo dehiscentes, & flexui etiam plures in digitis magna ex parte contigêre.

Terra, ut sunt coturnices, pernices, & gallinæ omnia.

Ræcipuæ differrētiæ ..., quæ ex ...nato gene- ..., uel quæ .... habent ..ā, sumun- tur à

Partibus auium utpote, ex

Rostris, siue ore, quod quibus rostrum uocatur, & ijs loco dentium, labiorum manusq; sed uario pro usu & auxilio inseruit. Rostra ossea, & maiori ex parte cornu obduc̄ta sunt, atq; maximum in capitibus discrimē adferunt. Variant pro ratione uictus, alijs enim concessum est

Rectum, utpote ijs, quæ cibi capiendi gratia id tantum habent.

Acutum, ut culticlegis.

Vncum, hæ sunt aut

Carniuorum. His enim rostrum uncum utile est ad retinendam id, quod raptu ceperunt, & prædam ex animalibus adipiscendam, simul & ad uim inferendam. Vis enim in ung : & rostro est sita. Huius generis sunt aquilæ, accipitres, Falcones, uultures, milui, bubones & similes.

Minimè, nimirum quæ frugibus corticibus uel folliculis clusis uescuntur, & arbores scandunt : adunco enim rostro fruges deglubare, & truncos arborum apprehendere queunt. Huius generis suspsittaci, loxia, uel curuirostra auis. Nam hæ aues sint carniuoræ, me latet. Cum aues rostra ad latera non mouent, consequuntur carniuoro rostrum aduncum admodum & acutum.

Falcatum, quod aut falcis, aut arcus speciem imitatur, sit autem dupliciter.

Sursum, ut in auosella apparet.

Deorsum, ut in arquata atq; falcinello, ceu numero.

Latum ijs, quibus uictus ratio placida, & ex herbis comparanda. Tale enim rostrum ad effodiendum pabulum, & ad euellendum, & ad tondendum commodius est. Harum auium plurimarum rostrorum extremo corneus quasi unguis adhæret, quo cibos apprehendunt. Dentium loco circa rostrum striæ asperiusculæ inueniuntur, præcipuè in anatibus, his enim serratum rostri extremē est: ita enim herbarum carptus, quo uiuunt, facilius agitur. Cygnus siue olor est rostro parum turbinato, & falcis instar dentatos dentes habet in rostro minutissimos.

Longum ut & collum, quoniam cibum ex alto capere solent : bona parte men earum, tum etiam palmipedum, aut simpliciter, aut parte eadem captura bestiolarum aquatilium uiuit. Itaq; fit, ut collo, quasi arundine piscatoria rostro, ueluti linea, & hamo utantur.

Robustum & prædurum, uerbi gratia roborisecorum & carniuororum generi tale contigit uo.

Lautum & mollius minuti generis auibus ad terræ fructus colligendos, & ad bestiolas capiendas idoneum.

Collo, quod eadem de causa, qua alijs animantibus porrectum est, & ferè pro crurum modo magna ex parte descriptum est. Vt quibus

Longa sunt crura, ijs collum longum, ut sunt aues in paludibus & palustribus locis degentes. His etiam in principio purpurā uel digitorum membrana reperitur, quæ securitatis causa digitorum principia coniungit, in solo præmoli & durodegunt.

Breuia habent crura, ijs collum breue palmipedibus exceptis.

Causa est

Quòd collum nec breuite cum cruribus longis, nec longum cum cruribus breuibus prosis ex terra administrare potest. Carniuoris longitudo colli incommoda est, imbecille enim quod prolixū est, carniuoris aut, uictus beneficio uirili comparatur: quamobrem nulli colli longi, cui ungues adunci.

Pedum digiti diuisi, parum diducti, & tamen membris uix iuncti sunt, atq; propterea natura palmipedum est, his longa sunt concessa colla, talia enim ad cibum ex humore petendum commodiores sunt. Breuia habent crura, quo melius natare possint. Huius generis sunt olores, siue cygni, anseres, & multi mergi.

Calcaribus quædam aues

Calcaria habent, quæ grauium nonnullis præsidio sunt, nam nice alarum suis cruribus calcaria gerunt. Huius generis sunt in gallinaceorum genere galli, sic & anseres scotici capricalei dicti.

Calcaribus tanquàm inutilibus uncungues priuantur.

Aliæ omnes aues calcaribus destitutæ sunt.

Pedibus, quibus donantur uita, quam degunt, ratione. Prima distinctio uoluerū, inquit Plinius, maximè constat pedibus. Ego hic ex Aristotele & Plinio differentias auium ex pedum diuersitate desumptas colligere institui. Aues hic sunt

Apodes, id est, pedibus priuatæ, quæ uolatu præcipua munia perficiunt, sunt aues paradyseæ pedibus omnino carentes, & hirundines quædam, quarum marinæ ob breuia crura, quibus donatæ sunt apodes maiores, Syluestres, pedes minores dictæ sunt.

Sunt fissipedes, qui digitos habent separatos, et nulla membrana uel cute coniunc̄tor. Hæ rursus obtinent uel

Digitos rectos expansarūq; , ut quæ iuxta terram, uel in syluis ex frugibus, uermiculis et similibus uiuunt. Hæ rursus obtinent uel

Ternos digitos anteriori in regione pedis, & unicum in posteriori, qui loco calcis existit, tales sunt gallinacei generis omnes aues, pernices, alauda, columba, & infinitæ aliæ.

Ternos digitos specie auis, quæ hoc digito prorsus carere uidentur. In uanellis, gruibus, uialis ex nanclorum specie auis, & omnibus quasi uolatilibus, quæ longa habent crura, & iuxta decorere, & gallinacei aquaticis & omnibus aquatilium uiuunt, postremum digitum admodum breuis, & parē quæ ex captura bestiolarum aquatilium uiuunt, postremum digitum impedimento foret: alijs uelocem cursus est. Alijs, forsan in paludibus & aquis hic digitus impedimento esset.

Binos digitos ante, & binos post, ut in hirundine saxatili, iynge, quod eius corpus minus, quàm cæterarum propensum est in aduersum, embriza, pratensi, pico maximo, uel nigro, pico uario & uiridi, psittaci. In his interni digiti sunt exterius breuiores.

Digitos curuos ob ungues aduncos, ut sunt uncungues aues. Harum aliæ

Ternos habent digitos ante, & unum retrò, ut sunt accipitres, falcones, & aquilæ buteo similes.

Oscines aues in caueis domi enutritæ etiam carniuoræ obtinent curuos ungues, minimè tamen sunt carniuoræ.

Binos antè, & binos post, ut in noctua siue ulula, tinnunculo, accipitre, aquila aquataria conspicitur.

Ad extremitatem digitorum pertingit.

Sunt palmides uel solipedes. Planos pedes consequuntur in tericc̄ta inter digitos membrana , uel digito in palmæ effigie contextis obtinent, ijs enim utuntur ad nandum, ut remis nautæ, & pinnulis suis pisces. Membrana in his uel

Integra & continua est, minimè diuisa , ut in omnibus ferè palmipedibus contingit. Rursus in his membrana uel

Coniungit tres anteriores digitos, quartum & posteriorem liberum relinquit, ut in auibus, quæ simul grādinntur et uolant & natant, quales sunt anates, anseres, olores & similes. In his rursus membrana uel

Simul coniungit posteriorem anterioribus , ut in carbone aquatico, qui ad nandum & uolandum, & non ad currendum natus esse uidetur.

Diuisa minimè, coniunc̄ta exsistit , huius generis sunt mergi aliquot præditi ternis tantum digitis pedum latis adhærentibus membranis, ad aliquousq; diuisis, non, ut in cæteris aquaticis coniunc̄ti. Hoc uidetur in columbis maioribus mergorum generi ascriptis. Habent hæ aues ungues alijs auium unguibus latiores, præsertim medium, & habent membratim utrinq; scissam, aut saltem ab utraq; parte iuxta articulorum digitos inprimis indium, Similes pedes habent trapezorolæ uel merguli. Est & huius generis sulica, cuius digitis pedum membranæ nigræ latæ adhærent.

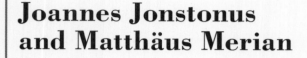

Joannes Jonstonus and Matthäus Merian

The Natural Histories

Joannes Jonstonus (1603–1675), *Historiae Naturalis de Quadrupedibus,* **1657; Birds and bats.**

Joannes Jonstonus was a physician and scientist of Scottish descent, born in Poland, who worked throughout northern Europe – and he was to set the standard for zoological literature for the next hundred years.

In the mid-seventeeth century, he embarked upon one of his many great projects: a linked natural history of insects, 'bloodless' marine animals, fish, whales and birds. His survey was wide-ranging and detailed, and benefited greatly from the striking contributions of Swiss engraver Matthäus Merian.

From a rich Basel family, Merian was not only an artist, but conveniently enough, also owned a publishing house. His engravings are remarkable in their accomplishment, but also their empathy for their subjects, who seem to wriggle and flutter from the page. Especially endearing is the female bat, whose tiny sucklings cling to her, as if to emphasize that bats are biologically very different from birds.

Merian died at the age of fifty-six, but his legacy was to prove even greater than his own artistry – his daughter Maria (see page 62) was to become, if anything, even more influential.

Opposite: Joannes Jonstonus (1603–1675), *Historiae Naturalis de Exanguibus Aquaticus,* **1657; Octopus and squid.**

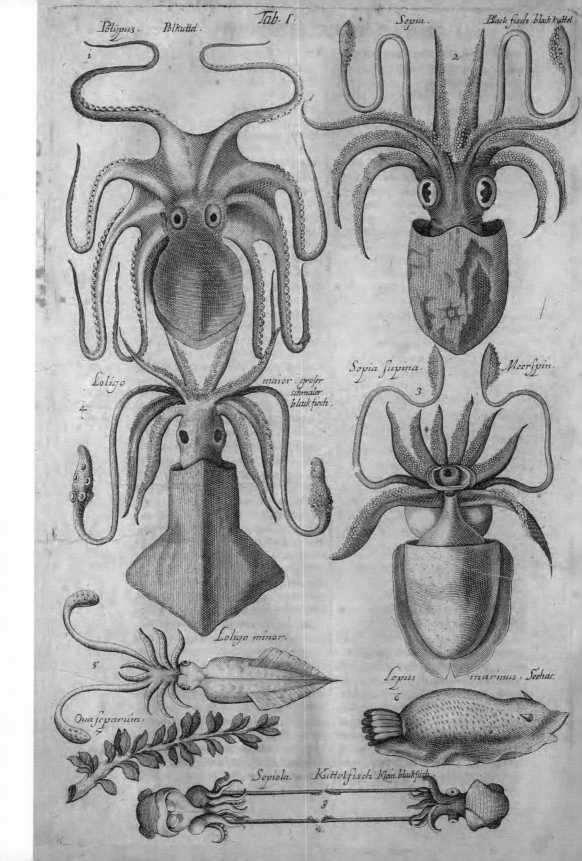

Tab. I.

Polypus. *Polkuttel.* ... *Sepia.* *Black fisch black kuttel.*

Sepia supina. *MeerSpin.*

Loligo *maior. groser schmaler black fisch.*

Loligo minor.

Lepus marinus. Seehas.

Oua separum.

Sepiola. *Kuttelfisch Klein blackfisch.*

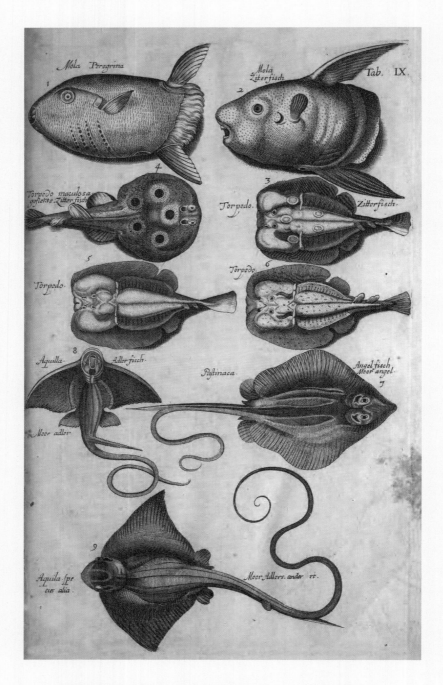

Above: Joannes Jonstonus (1603–1675), *Historiae Naturalis de Piscibus et Cetis*, 1657; Sunfish and rays.

Opposite: Joannes Jonstonus (1603–1675), *Historiae Naturalis de Quadripendbus*, 1657; Unicorns.

Tab. X

Monoceros Vnicornu
Einhorn

Capricornq̃ Marinq̃
Meer Steinbock

Monoceros Vnicornu
Einhorn

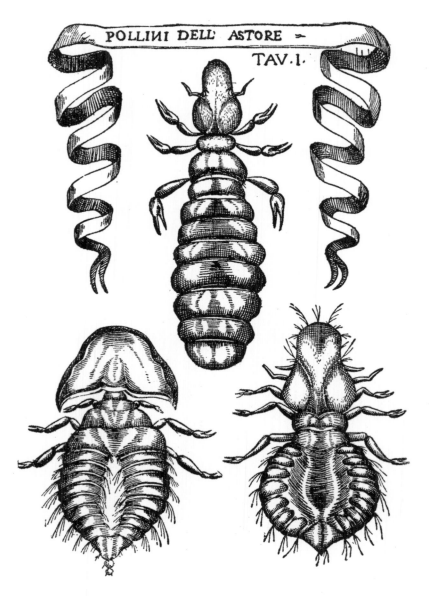

Francesco Redi (1626–1697), *Esperienze Intorno alla Generazione degli Insetti,* **1668; Lice.**

Redi was a scientist in the modern sense, who learned by observation and experiment. For example, he demonstrated that simple life forms do not appear spontaneously in decaying matter, but arise from eggs invisible to the naked eye. Only an enthusiastic entomologist would have included such a detailed view of lice in his book – 'Experiments on the Generation of Insects'.

Papio Parion 1   Papio 2   Mater Lamiarum clus

Arnoldus Montanus (1625–1683) and Jacob van Meurs (c.1617–1679),
De Nieuwe en Onbekende Weereld, **1671; Baboons and serpents.**

Neither the author (Montanus) nor the engraver (van Meurs) had visited the
'new' western and southern lands they present in this remarkable book ('The New
and Unknown World'), and some of the people and animals depicted presumably
strained the credulity of contemporary readers. Eagles attack unicorns, squid float
in mid-air and, in this example, an alarming crowd of varied beasts bursts from the
steamy jungle.

Left: John Ray (1627–1705) after
Francis Willughby (1635–1672),
Ornithology, 1678; Curlew,
curicaca, stilt and ibises.

The two naturalists Francis Willughby
and John Ray (see page 55) were long-
time collaborators, both interested in
the classification of animals and plants.
Ray developed Willughby's original
ideas into a system based on organisms'
similarities and differences. He is also
often credited as the originator of the
'species' concept.

Opposite: John Ray (1627–1705)
after Francis Willughby (1635–
1672), *De Historia Piscium*,
1686; Rays.

For the Royal Society, the financial
strain of publishing *De Historia Piscium*
was so great that it was unable to
fund publication of Isaac Newton's
Principia Mathematica.

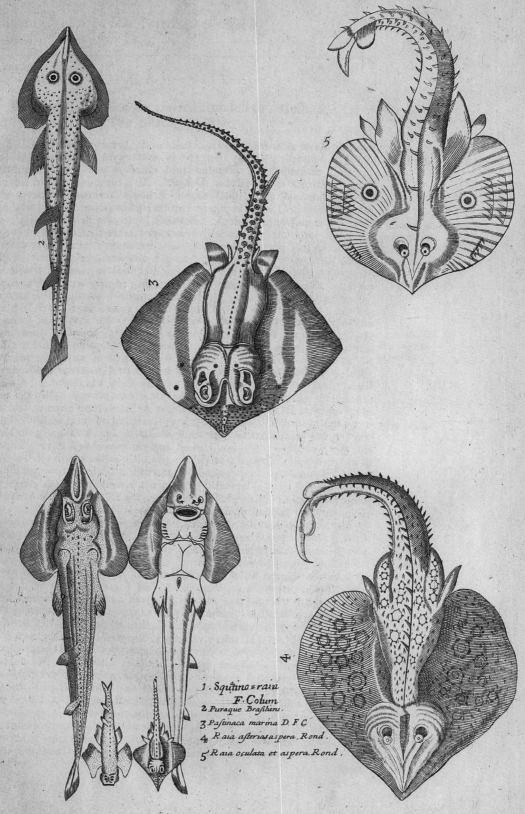

1. Squtino = raia
   F. Colum
2. Puraque Brasiliens.
3. Pastinaca marina D. F. C.
4. Raia asterias aspera. Rond.
5. Raia oculata et aspera. Rond.

Sumpt D. Samuelis Pepys Præsidis Societatis Regalis

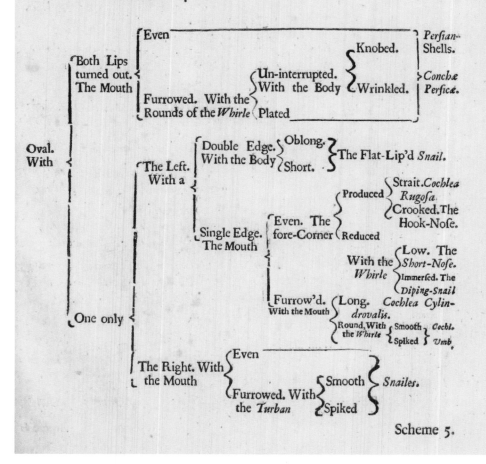

Scheme 4.

Scheme 5.

Nehemiah Grew (1641–1712), *Musaeum Regalis Societatis,* **1681;
Classificatory scheme of shelled animals (above); Gastrointestinal
anatomy of four mammals (opposite).**

Grew's work is part cabinet of curiosities and part enlightened analysis of the
biological world. Above is an obvious forerunner of modern species identification
keys, and opposite is a surprisingly schematic comparison of the alimentary
canals of four mammalian species, with half of the image dominated by the
torturous spirals, sacculations and speckled appendix of the humble rabbit.

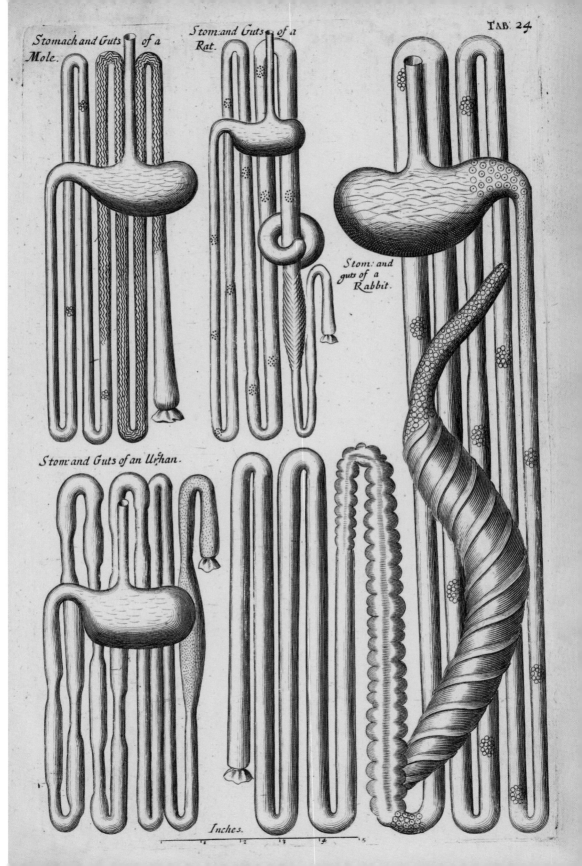

Stomach and Guts of a Mole.

Stom: and Guts of a Rat.

Stom: and guts of a Rabbit.

Stom: and Guts of an Urchan.

Inches.

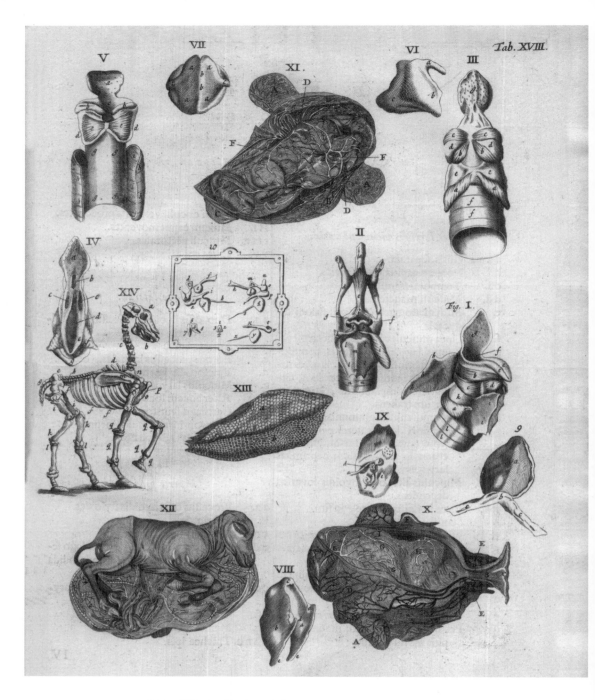

Gerard Blasius (1627–1682), *Anatome Animalium,* **1681; Horse anatomy, with some comparisons to humans.**

The prolific Dutch scientist Blasius published widely across the biological and physical sciences. In this image he draws upon comparative studies of equine structure largely carried out by anatomists in northern Italy.

Jan Luyken (1649–1712), *Twenty-eight heads of people and animals,* **1682.**

Apparently assembled for entirely artistic reasons, this montage by Dutch poet and engraver Jan Luyken plays to the idea of 'reverse anthropomorphism' – that people's superficial resemblance to animals might reflect deeper concordances in personality.

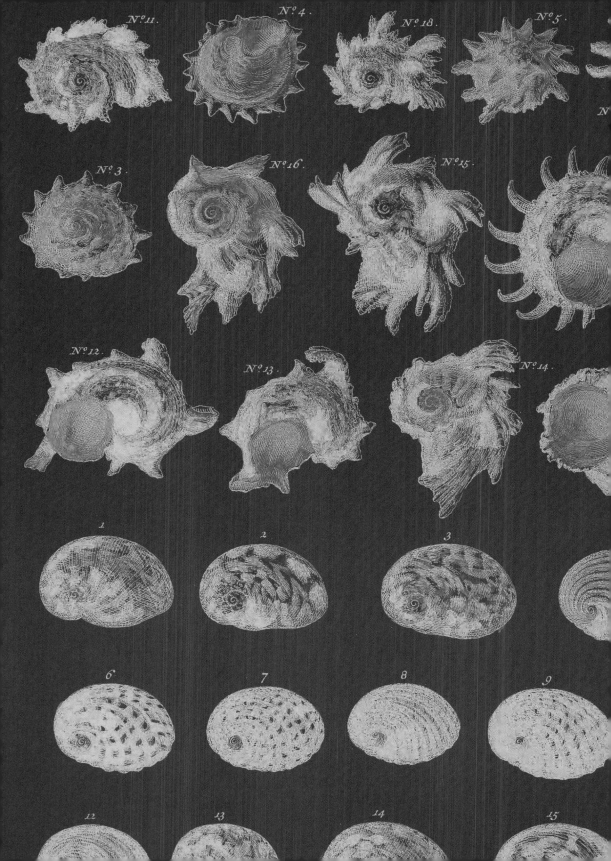

CHAPTER 2

Creating Order from Profusion

Albertus Seba (1665–1736), *Locupletissimi Rerum*, 1734–1765; Shells.

Renaissance and Enlightenment (1700–1820)

Our modern understanding of the relationships between animals is rooted in three concepts that were gradually formulated in the years between 1750 and 1870.

The three 'big ideas' are evolution, or the concept that animal species change, split and diversify over time; the ancient earth, the theory that enormous geological timespans have allowed evolution to create the animal complexity and diversity around us; and natural selection, the mechanism now thought to cause this evolutionary change.

These theories did not appear simultaneously as a complete, interlinked triad, but instead arose piecemeal as evidence accumulated and the intellectual shackles of the religious past were progressively cast off. An all-powerful God might have needed only a week to create the unchanging fixity of his animal kingdom, and only a few thousand years to play out the world's history, but by the Enlightenment many thinkers had realized that the Genesis narrative does not fit the evidence.

Maria Merian (1647–1717), *Metamorphosis Insectorum Surinamensium*, 1705 (see page 62); Spectacled caiman and false coral snake.

In retrospect, the eighteenth century was the pivotal time in the history of biology. Over a relatively brief period the seeds of doubt were sown in natural philosophers' minds about the origins of all the animal diversity around them. Throughout this period new discoveries reported from around the globe added to a growing sense that animal forms are almost endless in their variety, and an ever-accumulating fossil record showed the animal kingdom to be immensely variable not only in the present, but also across past aeons.

As a result, over the course of the eighteenth century, the emphasis of zoological classification changed from a confident cataloguing of the finite, discrete products of God's creation into an uncertain scramble to make sense of overwhelming animal variety. By the end of the period covered by this chapter, 1820, evolution and the ancient Earth were widely discussed – although natural selection would have to wait a little longer.

Maria Merian (1647–1717), *Caterpillars, Their Wondrous Transformation and Peculiar Nourishment from Flowers*, 1679; **The mulberry tree bears fruit.**

The idea of evolution is older than many realize, and some strikingly modern-sounding theories had already been proposed in scientific environments older, yet less theologically constrained, than eighteenth century Western Europe. Some ancient Greek philosophers suggested that life on Earth originally arose from non-living matter, that animal types change over time, and even that land animals are derived from aquatic ancestors. Furthermore, they speculated that fossils are remains of animals that met their demise in some ancient epoch, perhaps due to environmental cataclysms such as floods. Similarly, thinkers from the early Islamic world also proposed that animals change over time, and even that this was mediated by a 'struggle for survival' not unlike modern theories of natural selection. In addition, they suggested that past changes in animals had occurred in an impermanent world whose geography could change radically, so that what is now land might once have been sea, and vice versa.

In Western Europe a major step towards modern modes of animal classification was made in the late seventeenth century by John Ray, at

St Catharine's College in Cambridge, where the author of this book now works. Mainly remembered as the originator of the concept of a 'species', he also developed a methodical approach to the organization of living things according to their characteristics, and applied it across the plant and animal kingdoms. His neat nested lists of animal groupings do not imply an evolutionary process, and certainly cannot be interpreted as evolutionary trees, yet they reflect a newly objective form of scientific classification. For example, Ray argued on the basis of internal anatomy that whales are mammals and not fish, as many believed. He also attempted a comprehensive arrangement of insects based on their life cycles and mechanisms of metamorphosis.

Although more famous, the next phase of zoological organization could be argued to be conceptually retrogressive, even though it led to the development of a system of animal classification still in use today. The Swedish biologist Petri Artedi commenced this work in his posthumously published *Philosophia Ichthyologia* of 1738, when he developed a set of classificatory hierarchies akin to modern genera, families, orders and classes. His colleague Carl von Linné (often Latinized to 'Linnaeus') then extended Artedi's ideas and applied them to other animals, plants and even minerals. The result was the compendious 1735 *Systema Naturae* (see pages 65 and 67), in which he sought to arrange all living and mineral things into a coherent classificatory system. Although all its gradations of classification (apart from the species) were arbitrary inventions rather than real natural phenomena, its neat filing of the sprawl of nature was so satisfying that we still use it, or at least a variant of it. Linnaeus also introduced the convenient binomial system of naming organisms by which humans, for example, are snappily termed *Homo sapiens*. In fact, Linnaeus' dispassionate inclusion of humans in his scheme was itself one of his most scientifically important innovations.

The next major figure in this scientific revolution was Georges-Louis Leclerc, Comte de Buffon, a Frenchman who in the mid-eighteenth century pursued a wide range of intellectual enquiries in his compulsion to discover the origins and relatedness of life on Earth. He studied, not without controversy, the origin of the Earth, the origin of life from non-living matter, the cognitive abilities of animals, and the status of 'lost' species found in rocks but never encountered alive. He was also the first to subdivide the history of the Earth into a series of

John Ray (1627–1705),
Synopsis Methodica, **1693;**
General table of the animals.

prehistoric eras. Indeed, he thought contemporary physics and palaeontology had revealed the Earth to be at least 50,000–70,000 years old – incomprehensibly ancient to those who believed what the Bible told them. He also noted that fossil animals are frequently buried in locations far distant from where similar living creatures roam today, so conjectured that animals change over time and that it is migration that causes them to do so. He even proposed that animals' characteristics could be inherited by a non-spiritual mechanism, in response to exposure to new environments. As if that were not enough, Buffon was one of the first to create diagrams of interrelationships between animals. He formulated a 'network' of the similarities between dog breeds that appears very modern, although it does not look like an evolutionary tree – probably because, unlike wild species, dog breeds have not diverged to the point at which they can no longer interbreed. Long before genes or DNA, Buffon set the scene for modern biology.

Thomas Pennant (1726–1798), *British Zoology,* **1776; Fitchet (polecat) and martin.**

In the late eighteenth century, the drift towards evolution and the ancient Earth continued. In his 1766 *Elenchus Zoophytorum*, the Prussian zoologist Peter Simon Pallas argued that the interrelationships of living things could not be represented by a metaphysically idealized linear *scala*,

THE ZEBRA.

Thomas Bewick (1753–1828) *A General History of Quadrupeds,* **1790; The zebra.**

but that zoologists should instead use nets, or branching trees. He wrote of life as a tree trunk that had divided into two – plants and animals – with each hemi-trunk consisting of 'generic' species, giving rise to lateral twigs and branches of more specialized types. Disappointingly, it is not thought Pallas ever actually drew his tree.

Soon after, in 1785, James Hutton directly addressed the question of the age of our planet in his *Theory of the Earth*. He argued that most of the processes that shape the face of the planet are either slow, such as erosion and sedimentation, or infrequent, such as volcanism. Thus, for the Earth to have attained its present state it must be far older than several thousand years. This theme was later extended by the self-taught surveyor and geologist William Smith (see page 90) in his 1816 *Strata Identified by Organized Fossils*. Smith meticulously recorded the fossils found in sediments across Britain and realized that each rock layer or 'stratum' contains characteristic remains of its own distinctive, *different* fauna. He even discussed whether gaps in his fossil story represented simple incompleteness of the geological record, or ancient extinction events. The world was beginning to look more ancient than the human mind could easily contemplate.

The idea of a branching genealogical evolutionary tree (a 'phylogenetic' tree as it was named later in the eighteenth century) seems obvious now, yet biologists took decades to settle on it as their preferred classificatory format. This slowness probably reflects the fact that the now-ubiquitous tree shape implies things about animal origins and animal change that many were unwilling to accept.

Marcus Elieser Bloch (1723–1799), *Ichthyologie ou Histoire Naturelle des Poissons,* **1796; Carp (***Cyprinus***) species.**

In fact, the first biological 'classificatory tree' was of plants, not animals. Augustin Augier's 'Arbre Botanique' of 1801 (see page 84) was a systematic attempt to compare plants according to clearly defined criteria – although it was still not meant to demonstrate genealogy, nor even changes in plant forms over time. However, the tree motif was not solely decorative, and Augier gave much thought to whether plants share certain features because those features are directly equivalent between different species (akin to what we now call 'homology'), or simply because species need similar structures to perform similar functions ('analogy'). Thus, Augier pondered some subtle evolutionary issues, without even admitting that evolution occurs – and also hinted that a similar pictorial approach could be used for animals.

Eight years later, the first phylogenetic 'tree' of animals appeared – in subdued form – in Jean-Baptiste Lamarck's *Philosophie Zoologique* (see page 87). Unlike Augier, Lamarck strongly believed that the patterns of organismal variation he observed could only be explained by evolution, so his trees were certainly meant to represent species change and species-splitting over time. He persisted with his trees and the later ones acquired more branches, and also more certainty, as tentative dotted lines were replaced by decisive black lines and braces. Unfortunately, Lamarck is now widely ridiculed for suggesting evolutionary processes that seem unlikely to modern ears. Although he correctly stated that the environment is an important driver of animal evolution, he thought it acts by changing the extent to which individual animals use different organs and tissues, and that this relative use or disuse of parts of the body could be inherited by offspring.

He also clung to the old idea that there is an ideal of natural perfection towards which all organisms strive, even though in his mind this was by evolution rather than the determination of God or individual creatures. However, it is now often argued that Lamarck's errors are unimportant compared to his espousal of the inheritance of characteristics by physical rather than spiritual means, and the profound effect his enlightened support of evolution was to have on biologists who came after him.

Georges Cuvier (1769–1832), *Le Règne Animal,* **1817; Skull of a cod.**

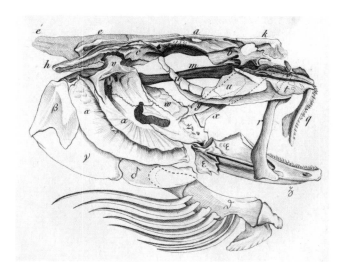

TAB: II.   pag. 204

F. 1.

F. 3.

F. 2.

Left: John Ray (1627–1705),
Three Physico-theological
***Discourses*, 1693; Fossils.**

Naturalist John Ray (see page 55)
corresponded extensively about the
nature of fossils, concluding they
were indeed the remains of dead
creatures, but he was uncertain
whether these creatures died due to
natural causes, or the biblical flood.
Any fossils with no known modern
equivalents were assumed to be
related to contemporary animals
as yet undiscovered.

**Opposite: William Derham
(1657–1735) and Eleazar
Albin (1690–1742),** *A Natural
History of English Insects***, 1720;
Hawk moths.**

Albin's chosen artistic medium was
watercolour and his usual subjects were
birds and insects. The *Natural History*
is a surprisingly accessible survey of
a topic of overwhelming size.

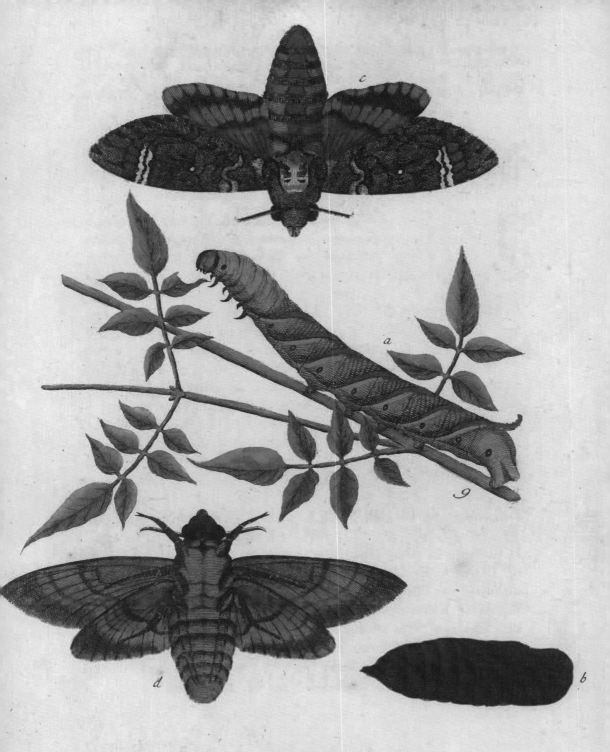

E. Albin. del.

TO THE RIGHT HONORABLE

Henrietta Somersett Countefs of Suffolk

this plate is humbly Dedicated by Eleazar Albin.

H. Fergusson. Sculp. London. 1713.

Maria Merian

Metamorphosis of the Insects of Surinam

Maria Merian (1647–1717), *Metamorphosis Insectorum Surinamensium*, 1705; Spiders.

Maria Sibylla Merian was an inspirational woman. The most gifted of a talented family, the daughter of Matthäus Merian (see page 40), she was trained mainly in the medium of watercolour, which was thought more seemly for a woman than oils or engravings.

Almost as soon as she started painting plants and flowers, Maria depicted the insects that devour them. Indeed, her early career shows a clear trend of the plants becoming less prolific, and the insects more so. Her most prominent devourers are caterpillars, and their metamorphosis into butterflies appealed to the devout Maria, who saw them as a symbol of the transformational relationship between humans and God.

However, Maria was also an able scientist, recording a variety of entomological discoveries, as well as providing strong support for Francesco Redi's recent refutation of the origin of insects from base matter by spontaneous generation (see page 44).

Having settled in the Netherlands, the enterprising Maria travelled with her daughter Dorothea to the tropical rainforests of Surinam, in South America, to indulge her passion for entomology. The results of their work was the stunning 1705 *Metamorphosis Insectorum Surinamensium*. Maria was clearly an astute businesswoman, as the book was made available in Dutch and Latin, in colour and monochrome, and also at reduced price to early purchasers.

Maria also wrote frankly and controversially about the conditions in the South American colonies – of how slaves employed suicide and toxin-induced abortions to avoid further suffering at the hands of the slaveholders.

Maria's relative social and scientific freedom stand in contrast to how her work was criticized and neglected in the nineteenth century. One famous image (left) shows a giant spider having killed a hummingbird, something many thought to be feminine hyperbole, but that we now know to be accurate.

Maria Merian (1647–1717),
*Metamorphosis Insectorum
Surinamensium*, 1705;
Butterflies and a caterpillar.

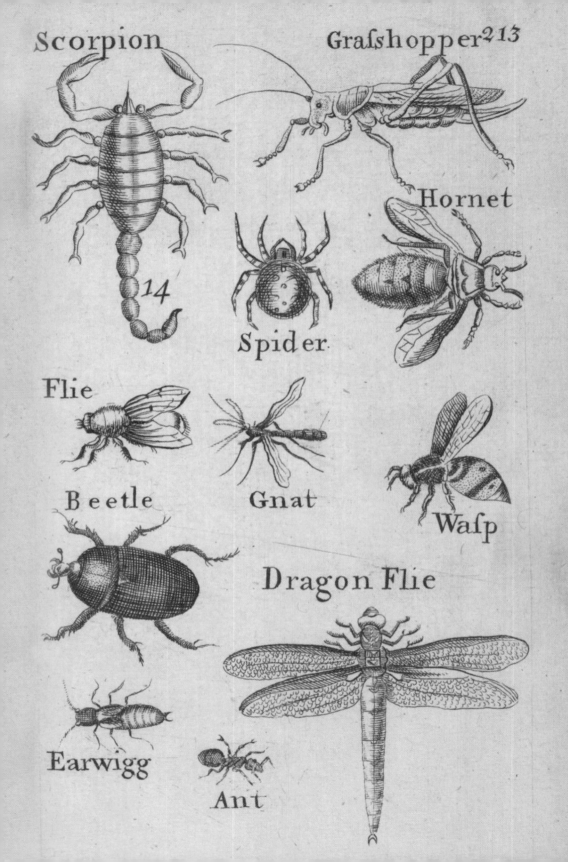

Scorpion

14

Hornet

Spider

Flie

Gnat

Beetle

Wasp

Dragon Flie

Earwigg

Ant

Above: Carl Linnaeus (1707–1778), *Systema Naturae*, **1735; Mineral kingdom.**

Published just a few months after he left medical school, *Systema Naturae* (see page 56) was Linnaeus' defining work. It introduced his new system of animal and plant classification, a scheme that was to be elaborated throughout the book's many subsequent editions. Linnaeus perhaps overstretched the concept when, as here, he endeavoured to include minerals in his scheme.

Opposite: Thomas Boreman (1712–1785), *A Description of Three Hundred Animals*, **1730; Insects and arachnids.**

Actually a children's book, the *Description* draws heavily on the bestiary tradition by mixing the familiar, exotic and mythical in its review of three hundred creatures. Boreman was an innovative publisher of children's books, and it is assumed he wrote this one.

CAROLI LINNÆI

I. QUADRUPEDIA.

Corpus hirsutum. *Pedes* quatuor. *Feminæ* viviparæ, lactiferæ.

Ordines.	Genera.	Characteres Generum.	Species.
ANTHROPOMORPHA. *Dentes primores* 4. utrinque; vel nulli.	Homo.	Nosce te ipsum.	H { Europæus albesc. Americanus rubesc. Asiaticus fuscus. Africanus nigr.
	Simia.	ANTERIORES. POSTERIORES. *Digiti* 5. . . . . . 5. Posteriores anterioribus similes.	Simia cauda carens. Papio. Satyrus. Cercopithecus. Cynocephalus.
	Bradypus.	*Digiti* 3. vel 2. . . 3.	Ai. *Ignavus.* Tardigradus.
FERÆ. *Dentes primores* 6. utrinque: *intermedii* longiores: *omnes* acuti. *Pedes* multifidi, unguiculati.	Ursus.	*Digiti* 5. . . . . . 5. Scandens. *Mamma* 4. (Ald.) *Calcaneis* insistit. *Pollex* extus positus.	Ursus. Coati *Mrg.* Wickhead *Angl.*
	Leo.	*Digiti* 5. . . . . . 4. Scandens. *Mammæ* 2. ventrales. *Lingua* aculeata.	Leo.
	Tigris.	*Digiti* 5. . . . . . 4. Scandens. *Mammæ* 4 umbilicales. *Lingua* aculeata.	Tigris. Panthera.
	Felis.	*Digiti* 5. . . . . . 4. Scandens. *Mammæ* 8. sc. 4. pect. 4. abdom. *Lingua* aculeata.	Felis. Catus. Lynx.
	Mustela.	*Digiti* 5. . . . . . 5. Scandens. *Dentes* molares 4. utrinque.	Martes. Zibellina. Viverra. Mustela. Putorius.
	Didelphis.	*Digiti* 5. . . . . . 5. *Mammæ* 8. intra bursulam abdomin.	Philander. *Possum.*
	Lutra.	*Digiti* 5. . . . . . 5. Palmipes.	Lutra.
	Odobænus.	*Digiti* 5. . . . . . 5. Palmipes. *Dentes* intermedii superiores longiss.	Ross. *Morsus.*
	Phoca.	*Digiti* 5. . . . . . 5. Palmipes. *Mammæ* duæ umbilicales.	Canis marinus.
	Hyæna.	*Digiti* 4. . . . . . 4. *Collum* superne jubatum. *Cauda* brevis.	Hyæna *Veter.* *Vivam Londini nuper vidit & descripsit* ARTED.
	Canis.	*Digiti* 5. . . . . . 4. *Mammæ* 10. sc. 4. pect. 6. abdom.	Canis. Lupus. Squillachi. Vulpes.
	Meles.	*Digiti* 5. . . . . . 5. *Ungues* medii digitis ipsis longiores. *Corpus* superne albicat: inferne nigricat.	Taxus. Zibetha.
	Talpa.	*Digiti* 5. . . . . . 5. *anteriores* maximi.	Talpa.
	Erinaceus.	*Digiti* 5. . . . . . 5. *Spinis* vel *lorica* squamosa munitus.	Echinus terrestris. Armadillo.
	Vespertilio.	*Digiti* 5. . . . . . 5. *Pes anticus* in alam expansus. *Mammæ* 2. pectorales.	Vespertilio. Felis volans *Seb.* Canis volans *Seb.* Glis volans *Seb.*
GLIRES *Dentes primores* 2. utrinque. *Pedes* multifidi.	Hystrix.	*Aures* humanæ. *Corpus* spinosum.	Hystrix.
	Sciurus.	*Digiti* 4. . . . . . 5. *Cauda* longissima lanigera.	Sciurus. . . . volans.
	Castor.	*Digiti* 5. . . . . . 5. palmipes postice. *Cauda* horizontalis, plana, nuda.	Fiber.
	Mus.	*Digiti* 4. . . . . . 5. *Cauda* teres, squamosa, hirsuta.	Rattus. Mus domesticus. . . brachiurus. . . macrourus. Lemures. Marmota.
	Lepus.	*Digiti* 5. . . . . . 4. *Cauda* brevissima, villosa.	Lepus. Cuniculus.
	Sorex.	*Digiti* 5. . . . . . 5. *Dentes canini* adsunt.	Sorex.
JUMENTA. *Dentes primores* incerti, obtusi. *canini* exerti, validi.	Equus.	*Mammæ* 2. inguinales. *Pedes* integri.	Equus. Asinus. Onager. Zebra.
	Hippopotamus.	*Mammæ* 2. inguinales (*Arist.*) *Pedes* quadrifidi.	Equus marinus.
	Elephas.	*Mammæ* 2. pectorales. *Pedes* 5. callis instructi.	Elephas. ? Rhinoceros.
	Sus.	*Mammæ* 10. abdominales. *Pedes* bisungulati: raro simplices.	Sus. Aper. Porcus. Barbyroussa. Tajacu.
PECORA. *Dentes primores* inferiores tantum: *Superiores* nulli. *Pedes* ungulati.	Camelus.	*Cornua* nulla.	Dromedarius. Bactrianus. Glama. Pacos.
	Cervus.	*Cornua* annua, primum pilosa, solida, ab apice crescentia. plurimis ramosa: feminis rara.	Camelopardalis. Caprea. Axis. Cervus. Platyceros. Rheno. *Rangifer.* Alces.
	Capra.	*Cornua* sursum versa. erecta, scabra.	Hircus. Ibex. Rupicapra. Strepsiceros. Gazella. Tragelaphus.
	Ovis.	*Cornua* retrorsum flexa intorta, rugosa.	Ovis vulgaris. . . Arabica. . . Africana. . . Angolensis.
	Bos.	*Cornua* antrorsum versa. lunulata, lævia.	Bos. Urus. Bison. Bubalus.
Ordines.	Genera.	Characteres Generum.	Species.

II. AVES.

Corpus plumosum. *Alæ* duæ. *Pedes* duo. *Rostrum* osseum. *Feminæ* oviparæ.

Ordines.	Genera.	Characteres Generum.	Species.
ACCIPITRES. *Rostrum* uncinatum.	Psittacus.	*Digiti* pedis antici 2. postici 2.	Psittacus.
	Strix.	*Digiti* pedis antici 3. posticus 1. quorum extimus retrorsum flexilis.	Bubo. Otus. Noctua. Ulula.
	Falco.	*Digiti* pedis antici 3. posticus 1.	Aquila. Vultur. Buteo. Falco. Cyanopus. Milvus. Lanius. Pygargus. Nisus. Tinnunculus.
PICÆ. *Rostrum* superne compressum, convexum.	Paradisæa.	*Pennæ* 2. longissimæ, singulares, nec alis, nec uropygio infixæ.	Manucodiata. Avis Paradisiaca.
	Coracias.	*Pes* 4dact. *Rectrices* exteriores gradatim breviores.	Pica.
	Corvus.	*Pes* 4dact. *Rectrices* æquales.	Corvus. Cornix. Monedula. Lupus. Glandaria. Caryocatactes.
	Cuculus.	*Digiti* pedis antici 2. postici 2. *Rostrum* læve.	Cuculus. Torquilla s. Junx.
	Picus.	*Digiti* pedis antici 2. *Rostrum* angulatum.	Picus niger. . . . viridis. . . . varius.
	Certhia.	*Pes* 4dact. *Rostr.* gracile incurvum.	Certhia.
	Sitta.	*Pes* 4dact. *Rostr.* triangulare.	Picus cinereus.
	Upupa.	*Pes* 4dact. *Caput* plumis cristatum.	Upupa.
	Ispida.	*Pes* 4dact. cujus digitus extimus medio adnectitur tribus articulis.	Ispida. Merops.
MACRORHYNCHÆ. *Rostr.* longiss. acut.	Grus.	*Caput* cristatum.	Grus.
	Ciconia.	*Ungues* plani, subrotundi.	Ciconia.
	Ardea.	*Unguis* medius inferne serratus.	Ardea.
ANSERES *Os* dentato-serratum.	Platelea.	*Rostr.* depresso-planum, *apice* suborb.	Platea.
	Pelecanus.	*Rostr.* depressum, *apice* unguiculato, inferne bursa instructum.	Onocrotalus.
	Cygnus.	*Rostr.* conico-convexum.	Olor. Elder. Anser. Ans. Bernicla.
	Anas.	*Rostr.* conico-depressum.	Anas fera. Glaucium. Boschas. *A. Domest.* Penelope. Querquedula.
	Mergus.	*Rostr.* cylindriforme, *apice* adunco.	Mergus. Merganser.
	Graculus.	*Rostr.* conicum, *apice* adunco.	Carbo aquat. Graculus aquat.
	Colymbus.	*Rostr.* subulatum. *Pedes* infra æquilibr.	Colymbus. C. minim. Podiceps. Arctica.
	Larus.	*Rostr.* subulatum. *Pedes* in æquilibr.	Cataracta. Larus. Sterna. Piscator.
SCOLOPACES. *Rostrum* cylindraceo-tereticulum.	Hæmatopus.	*Pes* 3dact. *Rostri* apex compressus.	Pica marina.
	Charadrius.	*Pes* 3dact. *Rostri* apex teres.	Pluvialis. Hiaticula.
	Vanellus.	*Pes* 4dact. *Rostrum* digitis brevius. *Caput* pennato cristatum.	Capella.
	Tringa.	*Pes* 4dact. *Rostrum* digitis brevius. *Caput* simplex.	Tringa. Ocrophus. Pugnax. Gallinula.
	Numenius.	*Pes* 4dact. *Rostrum* digitis longius.	Gallinago. Limosa. Arquata. Recurvirostra.
	Fulica.	*Pes* 4dact. *Digiti* membranis aucti. *Caput* carnoso-cristatum.	Gallinula aquatica.
GALLINÆ. *Rostrum* conico-incurvum.	Struthio.	*Pes* 2dact. absque postico.	Struthio-camelus.
	Casuarius.	*Pes* 3dact. absque postico. *Caput* galea & palearibus ornatum.	Emeu.
	Otis.	*Pes* 3dact. absque postico. *Caput* simplex.	Tarda.
	Pavo.	*Pes* 4dact. *Caput* corolla pennac. orn.	Pavo.
	Meleagris.	*Pes* 4dact. *Frons* papilla. *Gula* membrana unica longitudinali instructa.	Gallopavo.
	Gallina.	*Pes* 4dact. *Frons* membrana serrata. *Gula* membr. duplici longitud. inst.	Gallina.
	Tetrao.	*Pes* 4dact. *Supercilia* papillosa nuda.	Phasianus. Urogallus. Tetrao. Bonasia. Lagopus. Perdix. Coturnix.
PASSERES. *Rostrum* conico-attenuatum.	Columba.	*Rostr.* rectum, ad basin furfuraceum. *Nares* oblongæ, membranula superne instructæ.	Columba. Turtur. Palumbus. Oenas.
	Turdus.	*Rostr.* parum convexum. *Pluma* rostri basin tegunt.	Turdus. Merula.
	Sturnus.	*Rostr.* rectum subteres. *Lingua* bifida cornea.	Sturnus.
	Alauda.	*Unguis* digiti postici digitis ipsis longior.	Alauda.
	Motacilla.	*Rostr.* gracile. *Fauces* nigricant. *Lingua* apex bifidus laceratus.	Motacilla. Oenanthe. Merula aquatica.
	Luscinia.	*Rostr.* gracile rectum. *Lingua* apex bifidus laceratus.	Luscinia. Ficedula. Erithacus. Troglodytes. Carolina *Rudb.*
	Parus.	*Rostr.* gracile. *Lingua* apex truncatus, 4 setis instructus.	Parus. P. caudatus. . . . cristatus.
	Hirundo.	*Rostr.* gracile, ad basin depressum, minimum: *rictu* amplissimo.	Hirundo. Caprimulgus.
	Loxia.	*Rostr.* crassum, magnum, breve, curvum, undique convexum.	Coccothraustes. Loxia. Pyrrhula.
	Ampelis.	*Rostr.* crassum, rectum. *Remigum* apices nonnulli membranacei.	Garrulus Bohem.
	Fringilla.	*Rostr.* crassum. *Maxilla* utraque alteram sinu quodam ad basin recipit.	Fringilla. Carduelis. Emberiza. Spinus. Passer.

III. AMPHIBIA.

Corpus nudum, vel squamosum. *Dentes molares* nulli: *reliqui* semper. *Pinnæ* nullæ.

Ordines.	Genera.	Characteres Generum.	Species.
SERPENTIA.	Testudo.	*Corpus* quadrupedium, caudatum, testa munitum.	Testudo tessulata. . . . terrestris. . . . marina. Lutaria.
	Rana.	*Corpus* quadrupedium, cauda destitutum, squamis carens.	Bufo. Rana arborea. . . aquatica. . . Carolina.
	Lacerta.	*Corpus* quadrupedium, caudatum, squamosum.	Crocodilus. Allegator. Cordylus. Draco volans. Scincus. Salamandra aq. . . terrestris. Chamæleo. Seps. Senembi *Mrg.*
	Anguis.	*Corpus* apodum, teres, squamosum.	Vipera. Cæcilia. Aspis. Caudisona. Cobras de Cabelo. Anguis Æsculapii. Cenchris. Natrix. Hydrus.

AMPHIBIORUM Classem ulterius continuare noluit benignitas Creatoris; Ea enim si tot Generibus, quot reliquæ Animalium Classes comprehendunt, gauderet; vel si vera essent quæ de Draconibus, Basiliscis, ac ejusmodi monstris οἱ τερατολόγοι fabulantur, certè humanum genus terram inhabitare vix posset.

PARADOXA.

HYDRA corpore anguino, pedibus duobus, collis septem, & totidem capitibus, alarum expers, asservatur Hamburgi, similitudinem referens Hydræ Apocalypticæ à S. JOANNE CAP. XII. & XIII. descriptæ. Eaque tanquam veri animalis speciem plurimis præbuit, sed falso. Natura sibi semper similis plura capita in uno corpore nunquam produxit naturaliter. Fraudem & artificium, cum Ipsi vidimus, dentes Ferino-mustelini, ab Amphibiorum dentibus diversi, facillime detexerunt.

RANA-PISCIS s. RANÆ IN PISCEM METAMORPHOSIS valdè paradoxa est, quum Natura mutationem Generis unius in aliam diversæ Classis non admittat. Ranæ, ut Amphibia omnia, pulmonibus gaudent & ossibus spinosis. Pisces spinosi, loco pulmonum, branchiis instruuntur. Ergo legi Naturæ contraria foret hæc mutatio. Si enim piscis hic instructus est branchiis, erit diversus à Rana & Amphibiis. Si verò pulmones, erit Lacerta: nam toto cœlo à Chondropterygiis & Plagiuris differt.

MONOCEROS *Veterum*, corpore equino, pedibus ferinis, cornu recto, longo, spiraliter intorto, Pictorum figmentum est. MONODON *Artedi* ejusmodi cornu gerit, cæteris verò partibus multum differt.

PELECANUS rostro vulnus infligens femori suo, ut emanante sanguine sitim pullorum levet, fabulosè ab Iisdem traditur. Ansam fabulæ dedit saccus sub gula pendulus.

SATYRUS caudatus, hirsutus, barbatus, humanum referens corpus, gesticulationibus valde deditus, salacissimus, Simiæ species est, si unquam aliquis visus fuit. *Homines* quoque *Caudati*, de quibus recentiores peregrinatores multa narrant, ejusdem generis sunt.

BOROMETZ s. AGNUS SCYTHICUS plantis accensetur, & agno assimilatur; cui caulis alterius plantæ è terra erumpens umbilicum intrat; idemque sanguine præditus à feris devorari temerè dicitur. Est autem artificiose ex radicibus Filicinis Americanis compositus. Naturaliter autem est *Embryo Ovis* allegoricè descriptus, qui omnia data habet attributa.

PHOENIX, Avis species, cujus unicum in mundo individuum, & quæ decrepita ex ferali busto, quod sibi ex aromatibus struxerat, repuerascere fabulosè fertur, felicem subiturà prioris vitæ periodum. Est verò PALMA DACTYLIFERA. vid. *Kæmpf.*

BERNICLA s. ANSER SCOTICUS & CONCHA ANATIFERA è lignis putridis in mare abjectis nasci à Veteribus creditur. Sed fucum imposuit *Lepas* interaneis suis penniformibus, & modo adhærendi, quasi verus ille anser *Bernicla* indè oriretur.

DRACO corpore anguino, duobus pedibus, duabus alis, Vespertilionis instar, est *Lacerta alata*, vel *Raja* per artem monstrose ficta, & siccata.

AUTOMA MORTIS Horologii minimi sonitum edens in parietibus, est *Pediculus pulsatorius* dictus, qui ligna perforat, eaque inhabitat.

I
Corpus

PLAGIURI. *Cauda* horizontalis.

CHONDROPTERYGII. *Pinnæ* cartilagineæ.

BRANCHIOSTEGI. *Pinnæ* ossic. carentes. *Branch.* oss. & membran.

ACANTHOPTERYGII. *Pinnæ* osseæ, quarum quædam aculeatæ.

MALACOPTERYGII. *Pinnæ* osseæ, quæ omnes molles.

V. INSECTA.
Corpus crusta ossica cutis loco tectum. Caput antennis instructum.

VI. VERMES.
Corporis Musculi ab una parte basi cuidam solidæ affixi.

Carl Linnaeus (1707–1778), *Systema Naturae*, 1735; The animal kingdom.

The centrepiece of Linnaeus' obsessive filing of the living world, the *Systema Naturae* (see page 56), this table divides animals into quadrupeds, birds, amphibians, fish, insects and worms ('vermes'). The inclusion of 'paradoxa' hints, however, at an admission that the author was not quite able to produce a comprehensive organization of the beasts.

The Cock Macaw from Jamaica.

Opposite: Eleazar Albin (1690–1742), *A Natural History of Birds,* **Volume 2, 1734; Albin's macaw.**

Albin's macaw is known only from this one image, so is assumed to be extinct. Sadly, many New World psittacine species are known only from compendia created by eighteenth- and nineteenth-century zoologists.

Right: Charles Bonnet (1720–1793), *Traité d'Insectologie,* **1745; Notion of a scale of living beings.**

One of the most detailed expressions of the idea of the *scala naturae* (see page 16) appears in this entomological treatise by French naturalist-philosopher Charles Bonnet.

IDÉE D'UNE ÉCHELLE
DES ETRES NATURELS.

L'HOMME.
Orang-Outang.
Singe.
QUADRUPEDES.
Écureuil volant.
Chauveſouris.
Autruche.
OISEAUX.
Oiſeaux aquatiques.
Oiſeaux amphibies.
Poiſſons volans.
POISSONS.
Poiſſons rampans.
Anguilles.
Serpens d'eau.
SERPENS.
Limaces.
Limaçons.
COQUILLAGES.
Vers à tuyau.
Teignes.
INSECTES.
Gallinſectes.
Tænia, ou Solitaire.
Polypes.
Orties de Mer.
Senſitive.
PLANTES.
Lychens.
Moiſiſſures.
Champignons, Agarics.
Truffes.
Coraux & Coralloïdes.
Lithophytes.
Amianthe.
Talcs, Gyps, Sélénites.
Ardoiſes.
PIERRES.
Pierres figurées.

Georges-Louis Leclerc, Comte de Buffon (1707–1788), *Histoire Naturelle*, **1749;** *L'Unau* **(two-toed sloth).**

Sloths are strange in many ways, and one of these is their classificatory relationship to other 'placental' (i.e. live-bearing, non-marsupial) mammals. In many modern classifications they are considered part of a 'sister group' that split early from the lineage leading to most other mammals. Leclerc (see page 56) was a key figure of the eighteenth-century scientific revolution.

Albertus Seba (1665–1736), *Locupletissimi Rerum*, 1734–1765; Shells.

The Dutch zoologist Albertus Seba compiled one of the largest biological collections then in existence, and published detailed catalogues of most of his specimens, too.

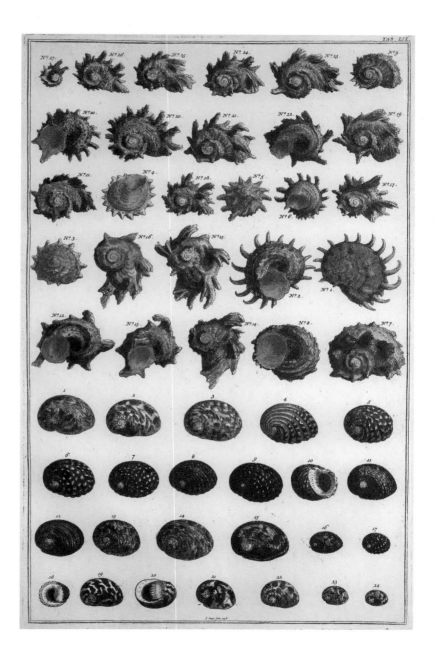

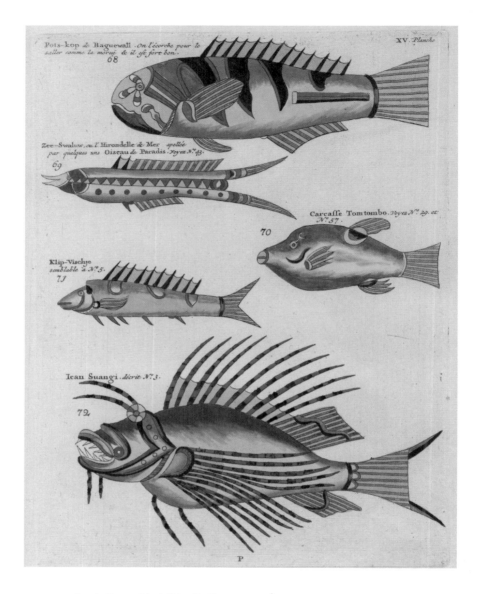

Louis Renard (c.1678–1746), *Poissons, Écrevisses et Crabes,* **1754;
Fishes (above); Crustaceans (opposite).**

Renard was an apothecary, publisher and spy, and also made time to be an
ichthyologist. The full title of this book translates as 'Fishes, crayfishes, crabs,
of diverse colours and extraordinary forms, found around the Molluccas
and on the coasts of the southern lands'. Some of the creatures depicted
are real, but many seem to have been made up, including a mermaid.
Also, their appearance becomes more fanciful as the book progresses.

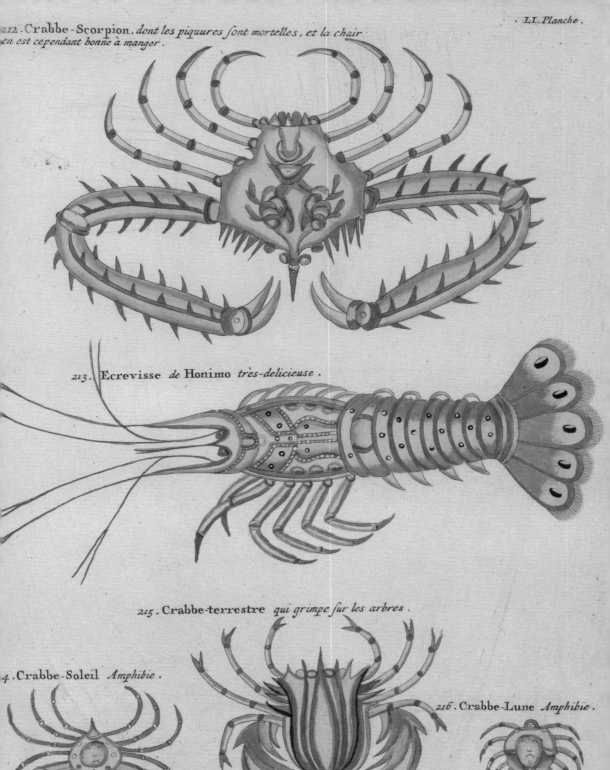

212. Crabbe-Scorpion. *dont les piquures sont mortelles, et la chair en est cependant bonne à manger.*

213. Ecrevisse *de Honimo très-delicieuse.*

215. Crabbe-terrestre *qui grimpe sur les arbres.*

4. Crabbe-Soleil *Amphibie.*

216. Crabbe-Lune *Amphibie.*

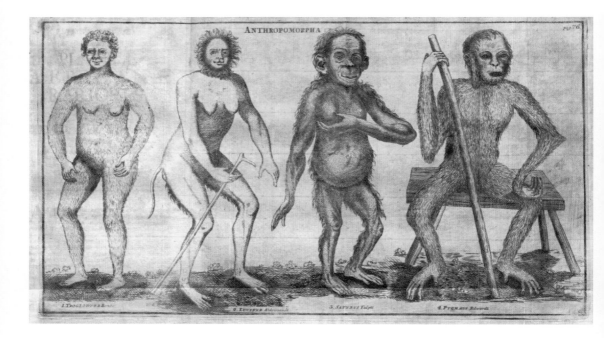

Above: Christianus Hoppius (dates unknown), *Linnaeus' Academic Delights,* **1763; Anthropomorpha.**

The origin of humans was a particular challenge to early anthropologists – partly because the lines between theology and science, myth and reality, and races and species had not yet been clearly drawn. This image, from a doctoral thesis published by Carl Linnaeus (see page 56), depicts *Troglodyta bontu, Lucifer aldovandri, Satyrus tulpii* and *Pygmaeus edwardi.* It is notable that the modern binomial name for the chimpanzee is still *Pan troglodytes.*

Opposite: Thomas Pennant (1726–1798), *British Zoology,* **1776; Creeper and hoopo(e).**

British Zoology was the most comprehensive catalogue of British fauna yet produced, but the cost of its beautiful engravings meant its author made no money from it. Pennant was much more than a cataloguer of animals, however, and corresponded widely with the natural philosophers of the day.

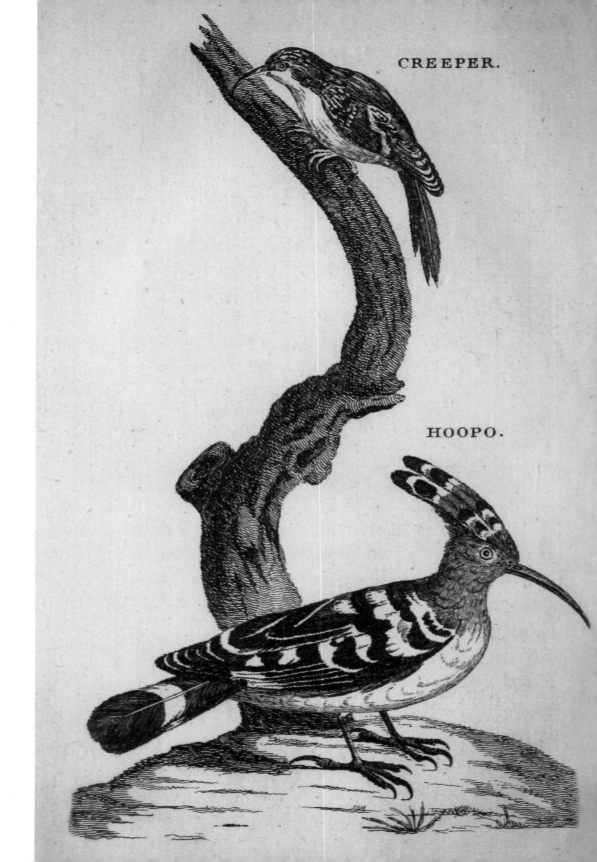

CREEPER.

HOOPO.

Above: Otto Friedrich Müller (1730–1784), *Zoologica Danica***, 1779; Whales.**

Only gradually were the affinities of whales elucidated. Even more than bats perhaps, cetaceans are the most dramatically specialized of all mammals. Although their ability to suckle their young had been known about for some time, many decades of diligent anatomical research were to pass until their origin was finally established – they are, in fact, most closely related to hippopotami.

Opposite: James Newton (1748–1804), *Dragonflies and mayflies* **(coloured engraving), 1780.**

Most insects that retain wings possess four of them, and nowhere is this as obvious as in the *Anisoptera*, the dragonflies and their allies. Fossil evidence of the group pre-dates that of dinosaurs by approximately 100 million years.

Ordo 4.
Insecta Neuroptera
GENUS I. *Libellulæ.*

GENUS II. *Ephemeræ.*

Ja.ᵗ Barbut del.

Ja.ˢ Newton Sculp.

Publish'd according to Act of Parliament Feb.ʸ 9:1780 by Sᵗ Barbut Nᵒ 101 Strand.

THE WALRUS, OR SEA-HORSE.

Thomas Bewick (1753–1828), *A General History of Quadrupeds*, **1790;
The walrus or sea-horse.**

Bewick's *Quadrupeds* is one of the most endearing zoological catalogues in existence,
partly because of the emotions and personalities with which each animal is imbued.
Some are proud, some wily, but the walrus is apparently the least cerebral.

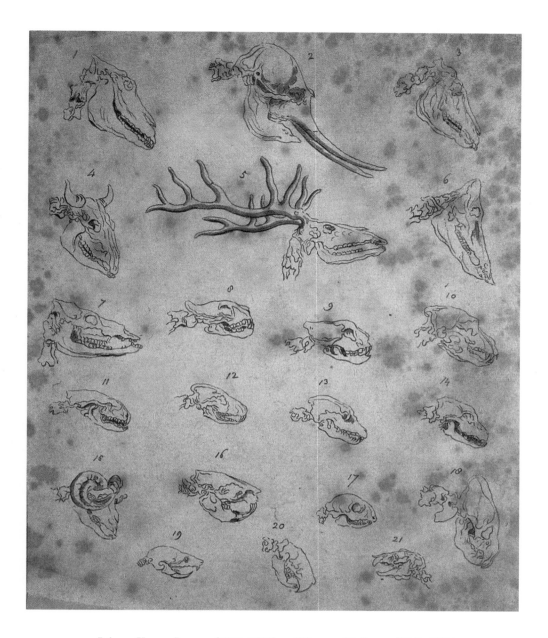

Johann Kaspar Lavater (1741–1801) and Thomas Holloway (1748–1827),
Essays on Physiognomy, **1789; Mammalian skulls.**

Lavater was a proponent of the field of physiognomy, the idea that a person's nature
and personality can be discerned from the way they look. In his studies, he drew
extensively from the facial structures of animals. Physiognomy was to become an
important element of common preconceptions and misconceptions about race.

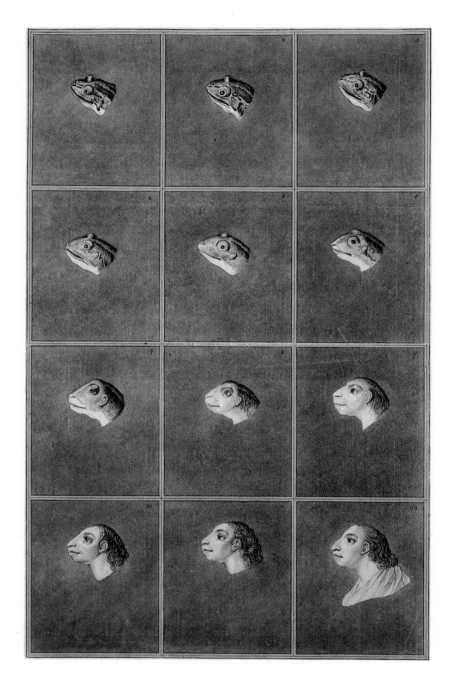

Johann Kaspar Lavater (1741–1801) and Christian von Mechel (1737–1817), *Sequence from the head of a frog to the head of a primitive man* (coloured etching), 1797.

Johann Kaspar Lavater (1741–1801) and Christian von Mechel (1737–1817), *Sequence from the head of a primitive man to the head of Apollo Belvedere* (coloured etching), 1797.

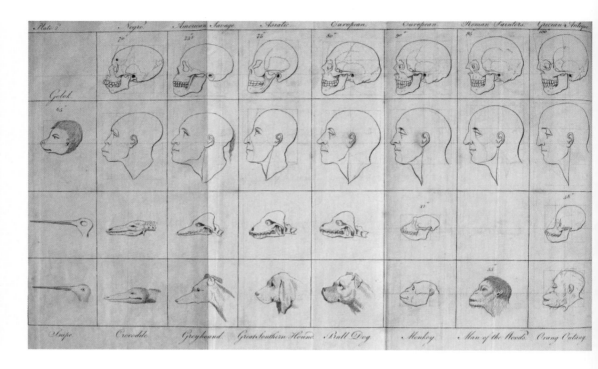

Charles White (1728–1813), *An Account of the Regular Gradation in Man, and in Different Animals and Vegetables,* **1799; The facial line of man, and different animals (above); Four different kinds of apes that approach nearest to man (opposite).**

White was an English physician whose urge to support the old idea of the *scala naturae* or 'great chain of being' (see page 16) can make for unpleasant reading today. In the image above, he ranks animals and humans according to the verticality of their facial profile, from snipe and crocodiles, via dogs and non-European humans, to what he describes as 'the model of the Graecian Antiques'. In the montage shown opposite are included 'Dr Tyson's pygmy' (presumably this is an orang-utan), 'a monkey', a 'native of Botany Bay' and 'an African' – the latter two compared with the profile of 'a European'.

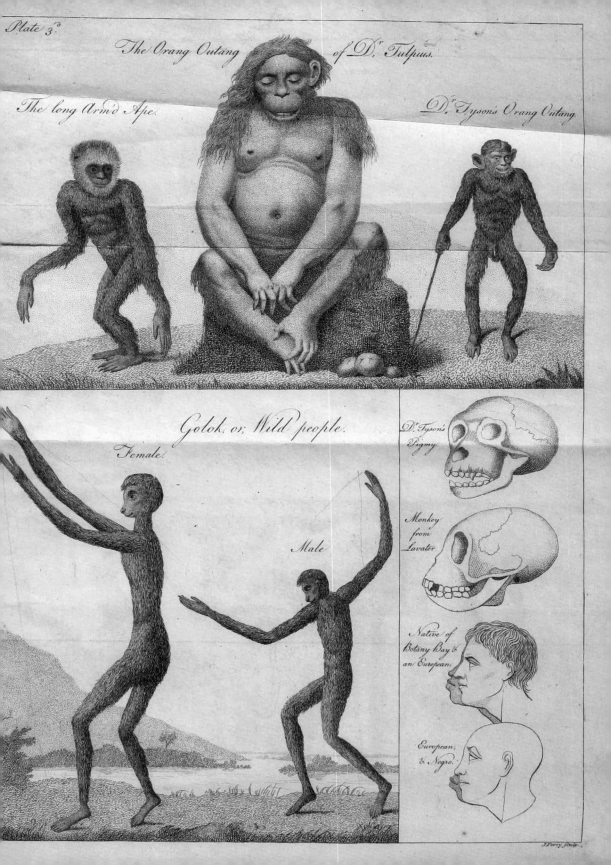

Plate 3.

The Orang Outang of Dr. Tulpius.

The long Arm'd Ape.

Dr. Tyson's Orang Outang.

Golok, or, Wild people.

Female.

Male.

Dr. Tyson's Pigmy.

Monkey from Lavater.

Native of Botany Bay & an European.

European, & Negro.

J. Perry sculp.

ARBRE BOTANIQUE.

Above: Jean-Baptiste Lamarck (1744–1829), *Tableau Encyclopedique et Méthodique des Trois Règnes de la Nature,* **1791; Sea urchins.**

Lamarck (see page 86) is now best known for his support of the theory of evolution, but in his early career he was a more orthodox classifier of organisms.

Opposite: Augustin Augier (1758–1825), *Essai d'une Nouvelle Classification des Végétaux,* **1801; 'Arbre botanique'.**

Augier's 'botanical tree' was probably the first tree-shaped classification of living things (see page 59).

Jean-Baptiste Lamarck

The Revolution of Evolution

J ean-Baptiste Lamarck was one of many siblings born to an impoverished old noble French family in the mid-eighteenth century. At seventeen, Lamarck journeyed to the south of France where he enlisted as a soldier, but more important to posterity than his successful military career was the fascination he developed with the flora growing in the hot dusty soils of the south.

Lamarck subsequently embarked upon medical training, but was later encouraged to change career by his brother. He worked first at the Jardin des Plantes in Paris, where he published a weighty classification of the plants of France, in which the avid botanist-reader is encouraged to identify their specimens using dichotomous keys similar to those developed by John Ray (see page 55).

It was perhaps these forerunners of modern decision trees, as well as the vagaries of his career path, that led Lamarck to make his greatest contributions to biology (a term he himself coined).

When appointed a founding academic at the new Muséum National d'Histoire Naturelle, Lamarck found the major botany posts already filled by more esteemed professors, so he developed a new niche studying invertebrate animals. Intrigued by the striking similarities he observed between species, he soon became a supporter of the controversial idea that organisms change over time.

Lamarck was not the first to propose the idea of evolution, but his work was to be some of the most influential in the field. Daringly, he proposed that there are natural, physical forces that drive evolutionary change. Two forces, in fact: one inherent force that drives animals to attain higher levels of complexity over time, and a second environmental force that forces animal species to diversify from each other.

**Jean-Baptiste Lamarck
(1744–1829),** *Histoire
Naturelle des Animaux
Sans Vertèbres* **(1815–1822);
Presumed organization of
the formation of animals.**

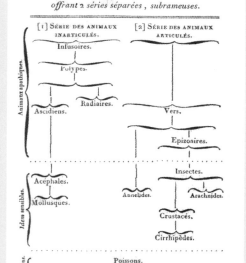

Jean-Baptiste Lamarck (1744–1829), *Philosophie Zoologique*, 1809; Chart showing the origins of different animals.

From Lamarck's most important book, *Philosophie Zoologique* (see page 59), this humble chart – 'Serving to Show the Origins of Different Animals' – is probably the first evolutionary tree of animals ever drawn. It is actually an inverted tree, with its root in the worms ('vers') at the top, and a selection of egg-laying, hoofed and marine mammals at the bottom. This dichotomously branching tree, focusing on just a few selected groups, is surprisingly prescient of the austere taxonomic diagrams of the late twentieth century.

ADDITIONS.   463

TABLEAU

Servant à montrer l'origine des différens animaux.

Vers.

Infusoires.
Polypes.
Radiaires.

Insectes.
Arachnides.
Crustacés.

Annelides.
Cirrhipèdes.
Mollusques.

Poissons.
Reptiles.

Oiseaux.

Monotrèmes.

M. Amphibies.

M. Cétacés.

M. Ongulés.

M. Onguiculés.

Cette série d'animaux commençant par deux

Sadly, Lamarck is now largely remembered for errors in the mechanisms he proposed – that simple life is continually spontaneously generated from inert matter, and that animals' life experiences can affect the morphology of their offspring.

These errors now seem forgivable, made as they were in the pre-genetic, pre-molecular age, and Lamarck now stands out as the first thinker to seek testable hypotheses about the mechanisms underlying animal diversity.

Alexander von Humboldt (1769–1859), *Ideen zu einer Geographie der Pflanzen,* **1807;** *Geographie der Pflanzen in den Tropen-Ländern.*

Although Humboldt studied mainly plants, he was influential in placing all organisms in their geographical context. Latitude, altitude, geographical proximity, meteorological variations – all affect which plants live where, and where plants live, so animals follow. Thus, Humboldt was a key figure in the origin of the modern science of biogeography (see Alfred Russel Wallace, Chapter 3, page 148). He also noted that the coastlines of some continents (for example, Africa and South America) are strangely complementary, and was thus a proponent of an early form of the theory of continental drift (see Alfred Wegener, Chapter 4, page 194).

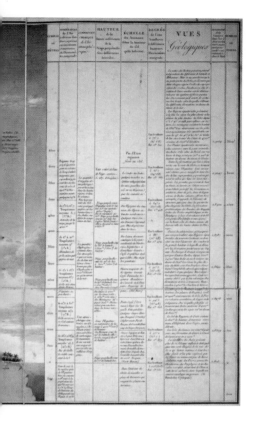

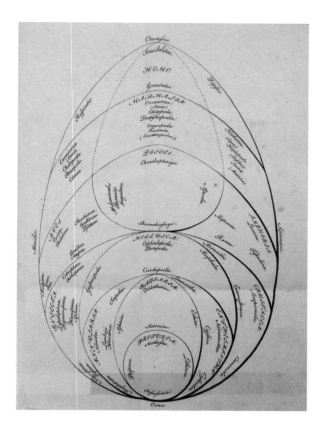

Georg August Goldfuss (1782–1848), *Über die Entwicklungsstufen des Thieres,* **1817; System of Animals.**

It sometimes appears that early-eighteenth-century biologists would go out of their way to avoid using a tree-like organizational scheme, with its controversial implications. Indeed, this egg-shaped diagram of animal classification by the Prussian palaeontologist Georg Goldfuss is an excellent example of such reticence. To the modern eye it resembles a Venn diagram, but is instead meant to depict similarities between animal groups, as well as an ascent to perfection – the capitalized words ascending the central axis are *protozoa* (single-celled animals), *radiaia* (anemones, sea cucumbers and others), *mollusca*, *pisces* (fish), *mammalia* and *homo* (man).

Opposite: William Smith (1769–1839), *Strata Identified by Organized Fossils,* **1816–1819; Typical fossils found in the Greensand stratum.**

Below: William Smith (1769–1839), *A Delineation of the Strata of England and Wales, with part of Scotland,* **1815.**

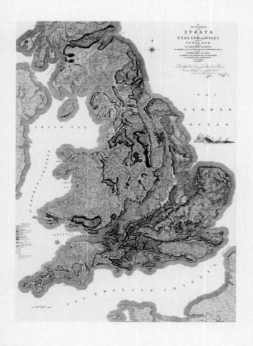

William Smith

The Principle of Faunal Succession

Today, palaeontology and evolutionary biology are usually assumed to be aspects of the biological sciences, yet they have their roots firmly in geology. This is because, in the nineteenth century, geologists realized that the chronology of animal evolution is recounted in the regularly deposited strata of rocks. In other words, layered deposits are the 'pages' of the story of animal life.

An important figure in this realization – one of the key realizations in the entire history of science, indeed – was William 'Strata' Smith.

Britain was the ideal place to study the layering (stratification) of ancient rock deposits, for two reasons. First, its small area fortuitously contains an almost complete series of diagonally aligned strata from some of Earth's oldest in northwest Scotland, to some of its youngest in England's populous southeast. Second, geological knowledge was at a premium in eighteenth-century Britain, as the Industrial Revolution drove the digging of canals and the excavation of mines.

Coming from humble beginnings, William Smith worked for much of his life as a surveyor, and his work gradually convinced him that not only do fossil animals appear in certain strata – his 'principle of faunal succession' – but conversely that the presence of certain 'diagnostic species' could even be used practically to determine when strata had been deposited.

In a few brief years, Smith published the first-ever national geological map, as well as two catalogues detailing the correspondences between mineral seams and their distinctive fossil captives: the *Strata Identified* and the *Stratigraphical System*.

Quite simply, our view of the world's history would never be the same again. Yet Smith never benefitted personally from his discoveries. His flurry of richly illustrated publications bankrupted him; he spent time in a debtor's prison, and died in poverty.

GREEN SAND.

1. *Serpula.*
2. *Solarium.*
3. *Pecten.*
4. *Terebratula pectinata. Min.Con.t.15.f.1.*
5. ——— *Lyra. Min.Con.t.138.f.2.*
6. *Terebratula.*
7. *Chama haliotidea. M.C.t.25.*
8. *Pecten 4-costata. Brit. Min.t.184.*
9. *Pecten.*
10. *Ostrea.*
11. *Echinus.*
12. *Echinites.*
13. *Echinus lapis cancri.*
14. *Spatangus.*
15. *Cycloliles. ban.*
16. *Madrepnile.*
17. *Alcyonite.*

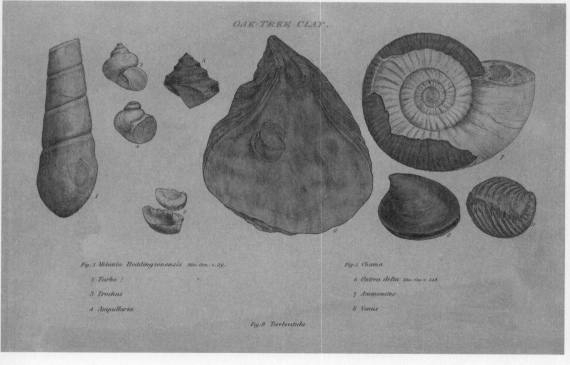

OAK-TREE CLAY.

Fig. 1 *Melania Heddingtonensis. Min.Con.t.39.*
2 *Turbo ?*
3 *Trochus*
4 *Ampullaria*

Fig. 5 *Chama*
6 *Ostrea delta. Min.Con.t.148.*
7 *Ammonites*
8 *Venus*

Fig. 9 *Terebratula*

GEOLOGICAL TABLE OF BRITISH …

WHICH IDENTIFY THE COURSES AND CONTINUITY OF THE STR…

AS ORIGINALLY DISCOVERED BY W. SMITH, Civil E…

GEOLOGICAL MAP OF ENGLA…

ORGANIZED FOSSILS which Identify the respective STRATA.		NAMES OF STRATA on the Shelves of the GEOLOGICAL COLLECTION	COLOURS on the MAP OF STRATA	NAMES in the MEMOIR and the PECU…
Volute, Rostellaria, Fusus, Cerithia, Nautili, Teredo, Crabs Teeth, and Bones	Plains	London Clay	Nº 1	London Clay forming Highgate, Harrow, Shoot…
Murices, Turbo, Pectunculus, Cardia, Venus, Ostreæ	Plains	Crag — Sand / Sand	2 / 3 / 4	Clay or Brickearth with Interspersions of Sand… / Sand & light Loam upon a Sandy or absorbent
Flint, Alcyonia, Ostreæ, Echini, Plagiostoma	Chalk Hills	Chalk Upper / Lower	5	Chalk Upper part soft contains flints / Lower part hard contains none
Terebratula, Teeth, Palates, Plagiostoma	Chalk Hills			
Funnel form Alcyonia, Venus, Chama, Pectines, Terebratula, Echini	Chalk Hills	Green Sand	6	Green Sand parallel to the Chalk
Belemnites, Ammonites	Clay Vales	Brickearth	7	Blue Marl
Turritella, Ammonites, Trigoniæ, Pecten, Wood	Clay Vales	Portland Rock — Sand	8 / 9	Purbeck Stone Kentish Rag and Limestone of Pickering and Aylesbury
Trochus, Nautilus, Ammonites in Masses, Ostreæ (in a bed) Bones	Clay Vales	Oaktree Clay — Sand	10	
Various Madrepore, Melania, Ostreæ, Echini, and Spines	Clay Vales	Coral Rag and Pisolite	11	Iron Sand & Carstone which in Surry and Bedf… / Fullers Earth and in some Places Ochre
Belemnites, Ammonites, Ostreæ	Clay Vales	Clunch Clay and Shale — Sand	12 / 13	Dark blue Shale producing a strong Clay Soil ch… / in North Wilts and Vale of Bedf…
Ammonites, Ostreæ	Clay Vales	Kelloways Stone	14	
Modiola, Cardia, Ostreæ, Avicula, Terebratulæ	Stonebrash Hills	Cornbrash	16	Cornbrash A thin Rock of Limestone chiefly aro…
	Stonebrash Hills	Sand & Sandstone	17	
Pectines, Teeth and Bones, Wood	Stonebrash Hills	Forest Marble	18	Forest Marble Rock thin Beds used for rough Pav…
Pear Encrinus, Terebratulæ, Ostreæ	Stonebrash Hills	Clay over the Upper Oolite	19	Great Oolite Rock which produces the Bath Freeste…
Madrepore	Stonebrash Hills	Upper Oolite	20	
Modiolæ, Cardia	Stonebrash Hills	Fullers Earth & Rock	21	
Madrepore, Trochi, Nautilus, Ammonites, Pecten	Stonebrash Hills	Under Oolite	22	Under Oolite of the Vicinity of Bath and the m…
Ammonites, Belemnites as in the under Oolite	Stonebrash Hills	Sand	23	
Numerous Ammonites	Stonebrash Hills	Marlstone	24	
Belemnites, Ammonites in mats	Marl Vales	Blue Marl	25	Blue Marl under the best Pastures of the midlan…
Pentacrini, Numerous Ammonites, Plagiostoma, Ostreæ, Bones	Marl Vales	Lias	26 / 27	Blue Lias / White Lias
	Marl Vales	Red Marl	28	Red Marl and Gypsum soft Sandstones and Sea…
Madrepore, Encrini in Masses, Producti	Coal tract	Redland Limestone	29	Magnesian Limestone / Soft Sandstone
Numerous Vegetables, Ferns lying over the Coal	Coal tract	Coal Measures	30	Coal Districts and the Rocks & Clays which acco… / generally a Sandstone beneath
Madrepore, Encrini in Masses, Producti Trilobites	Mountainous	Mountain Limestone	31	Derbyshire Limestone or Metalliferous Limestone
	Mountainous	Red Rhab & Dunstone	32	Red & Dunstone of the Southern and Northern Par… / Interspersions of Limestone marke…
	Mountainous	Killas	33	Various / Killas or Slate and other Strata of the Mountains / West Side of the Island with Inters… / of Limestone marked blue
	Mountainous	Granite, Sienite & Gneiss	34	Granite Sienite and Gneiss

From the re-examination of the Authors numerous Specimens in the arrangement of his Geological Collection in the British Museum and his subsequent observations this list of the St…

NIZED FOSSILS,

ORDER OF SUPERPOSITION;

REFERENCE TO HIS

VALES.

STRATA.   PRODUCTS of the STRATA.

ills

Septarium from which Parker's Roman Cement is made

{ No Building Stone in all this extensive District but Abundance
{ of Materials which make the best Bricks and Tiles in the Island

{ Potters Clay, Glass Grinders Sand, and Loam and Sands used for
{ Various Purposes

Flints the best Road Materials

Good Lime for Water Cements

Firestone and other soft Stone sometimes used for Building

The first Quarry and building Stone downward in the Series
Kimmeridge Coal

{ Fullers Earth, Ochre and Glass Sand
{   Some Lime used on these Sands in Sussex and Yorkshire

Makes tolerable Roads

Coarse Marble, rough Paving and Slate

{ The finest Building Stone in the Island for Gothic and other
{   Architecture which requires nice Workmanship

Excellent Lime for Water Cements

Now used for printing from M. S. written on the Stone

Small Quantities of Copper and Lead and Calamine

{ Grindstones, Millstones, Pavingstone, Iron-Stone and Fire-Clay
{ from the Coal Districts

Lead, Copper, Calamine   Marble

Some good Building Stone

The Limestone polished for Marble
Tin, Copper, Lead and other Minerals
{ The most durable building Stone in the Island for Bridges
{ and other heavy Works

and his future exertions will be in proportion to the encouragement which he receives from the Public.

William Smith (1769–1839),
Stratigraphical System of Organized
***Fossils,* 1817; Geological table of**
British organized fossils.

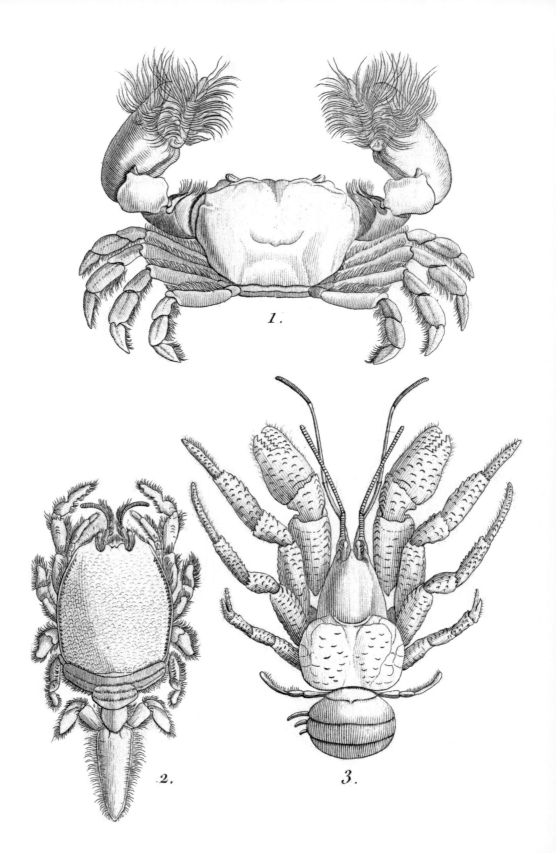

1.

2.

3.

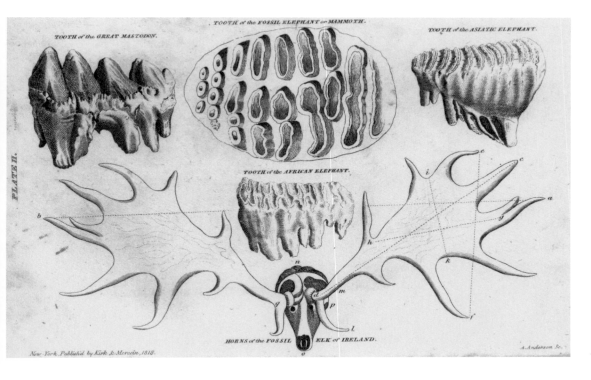

Georges Cuvier (1769–1832), *Le Règne Animal,* **1817; Crustaceans (left); and** *Théorie de la Terre,* **1817; Teeth of the extinct mastodon, African elephant and Asian elephant, and the horn (antlers) of the fossil elk of Ireland (above).**

Jean Léopold Nicolas Frédéric, Baron Cuvier, was an important figure in the development of modern modes of zoological organization. He expanded the classificatory system of Carl Linnaeus (see page 56), putting emphasis on the grouping of animals into large, overarching taxonomic groups. Importantly, he also incorporated fossil species into his system, and expounded the theory that periodic cataclysmic extinction events were important in the history of animal life. However, he did not support the developing theory of evolution.

Georges Cuvier (1769–1832), *Le Règne Animal*, 1817; Lizard, snake and frog (above); Birds (opposite).

Pierron sculp.

Trees of Life in a Newly Ancient World

After Evolution (1820–1900)

The turning point in the story of zoological classification came with the theory of natural selection – formulated in the middle of the nineteenth century, simultaneously by two British biologists, Charles Darwin and Alfred Russel Wallace.

Natural selection is a satisfyingly but deceptively simple theory: populations of animals exhibit variation in their characteristics; individuals inherit traits from their parents; and if traits confer an advantage on offspring's ability to thrive and reproduce, then those traits will be preferentially inherited by subsequent generations. Thus, over many years, species gradually develop new traits and *change*, and sometimes split into daughter-species ('speciate'), as well. Remarkably, almost all the diversity and complexity of life is now thought to be built upon this process.

The decades before and after the expounding of this theory produced some of the most dramatic visual depictions of animal organization ever created. The images were compelling before natural selection as biologists

Ernst Haeckel (1834–1919), *Natürliche Schöpfungsgeschichte*, 1868; Single or monophyletic pedigree of the stem of the backboned animals based on palaeontology.

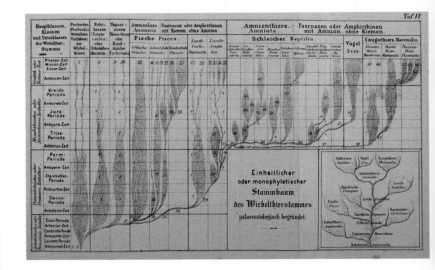

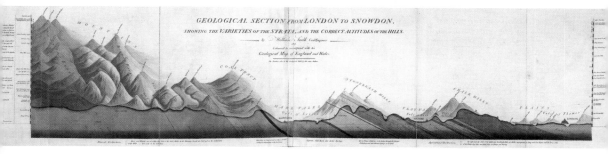

fought to make sense of an as-yet incomplete model of how the animal world works, and the images were also compelling afterwards as scientists realized how powerful the new biology could be.

In the decades after Lamarck's first evolutionary 'trees' (see page 86), other scientists struggled in their attempts to visually represent their proposed classifications of animals. Unable to cast off the young Earth or the unchanging creatures of the Bible, many went to great lengths to create elaborate, pleasingly geometric diagrams that nonetheless merely claimed to illustrate 'similarities', 'affinities' or the even more cryptic 'transits' between animal species. Some looked like exuberant Venn diagrams, some like mysteriously subdivided eggs and others like obscure Kabbalistic Sephiroth. Although inventive and visually appealing, it is hard to see why each of their creators felt so certain that their own personal geometry of animal relationships reflected nature's true scheme any better than anyone else's. To modern eyes it can seem that these biologists were doing all they could to avoid drawing what is to us (but not to them, of course) the obvious form – a tree.

During the first half of the nineteenth century, arguments continued to mount in favour of the Earth being much older than stipulated by the Bible narrative. This idea was to be very important for biologists because evolution is probably usually a slow process, so the theory of evolution 'needs' an old Earth to be credible. Although not himself the originator of the old-Earth hypothesis, the Scottish geologist Charles Lyell is often viewed as the central character in this geologico-biological concordance. Lyell was a proponent of 'uniformitarianism' – the hypothesis that geological conditions, processes,

William Smith (1769–1839), *Geological section from London to Snowdon* (**map**)**, 1817.**

Edward Hitchcock (1793–1864), *Elementary Geology,* **1840 (see page 114); Distribution of several orders of Zoophyta.**

SYSTEMS.	Sponge.	Lamellifere.	Crinoidea.	Echinida.	Stelerida.
Tertiary.	*	*	*	*	*
Cretaceous.	*	*	*	*	*
Oolitic.	*	*	*	*	*
Saliferous.			*		*
Carboniferous.					
Silurian.	*		*		*
Lower Systems.		*			

John James Audubon (1785–1851), *The Birds of America*, 1827–1838; Great blue heron.

and the forces that cause them, have been largely similar for extremely long periods of time. In his 1830-1833 *Principles of Geology*, Lyell described his own fieldwork showing that vertically stacked miles of sedimentary strata, and the accumulated conical layers of vast volcanoes, could only have formed over achingly long periods of time, and in accordance with unchanging laws of nature. A copy of the *Principles* was given to Darwin as he embarked on the *Beagle*, and Lyell's model of steady, cumulative geological change was to influence deeply the younger man's ideas about animals.

We now think the Earth is approximately 4.5 billion years old, and has supported life for at least eighty per cent of that time. However, the question of whether the ancient Earth was relatively unchanging or was instead racked by catastrophes has troubled geologists ever since the eighteenth century. Clear evidence of dramatic climate fluctuations was demonstrated by the discovery of past ice ages by the Swiss geologist Louis Agassiz in his 1840 *Études sur les Glaciers*. Glaciers cut distinctively U-shaped valleys and also transport and deposit 'erratics' – large lumps of rock – in the 'wrong place' (for example, huge granite boulders perched on top of younger limestone). Agassiz argued that glaciers had once reached as far south as the Caspian and Mediterranean seas, and in 1846 moved to the United States to bolster his theories. Once there, however, his career took an unexpected turn when he became embroiled in arguments about the nature of species, inter-species hybridization and the origin of the human races. He supported the theory of polygenism, the idea that the different human races actually represent different species – and thus is mainly remembered for supporting a theory widely used as a justification for slavery (see *Josiah Nott*, page 126, and his work with George Gliddon, pages 127–129).

All the while, the tradition of visually cataloguing animals continued to flourish into the nineteenth century. New territories and new research methods revealed ever more animal diversity to the artistically minded naturalist, and two particular works stand out. The first is John James

Audubon's colossal *The Birds of America*: published between 1827 and 1838, it contains 435 watercolours exhaustively and lovingly cataloguing the avian inhabitants of the new country. Contributing to these tomes' weightiness is the fact that Audubon decided to depict his subjects life-size, meaning that each page must be large enough to comfortably accommodate an eagle. The second major work is John Gould and Henry Constantine Richter's 1845–1863 *The Mammals of Australia* (see pages 140 and 141), which exposed European and American readers to the diversity of antipodean marsupial and egg-laying beasts. While not as huge as Audubon's work, *Mammals* shows an immensely endearing artistic empathy with its inhabitants. Rarely have such strange creatures looked so proud, furtive or playful. Some of them even appear to have a forlorn resignation in their eye, as if they know that, within a century and a half, some of them will no longer exist.

Throughout the early-to-mid-nineteenth centuries, scientific data accumulated at an unparalleled rate. In 1823, human and mammoth bones were discovered in a Welsh cave, suggesting the two species had coexisted. In 1844, the journalist Robert Chambers anonymously published his controversial *Vestiges of a Natural Creation*, in which he discussed the formation of the Solar System, the origins of terrestrial life, including humans, and how animal species may 'branch' off a stem of ever-increasing developmental complexity. In 1862, Lord Kelvin used the rate of thermal cooling of the Earth to calculate its age to be 20–100 million years, although even these enormous numbers did not satisfy

Louis Agassiz (1807–1873) and Augustus Gould (1805–1866), *Principles of Zoölogy*, 1851; Skeleton of a mammoth.

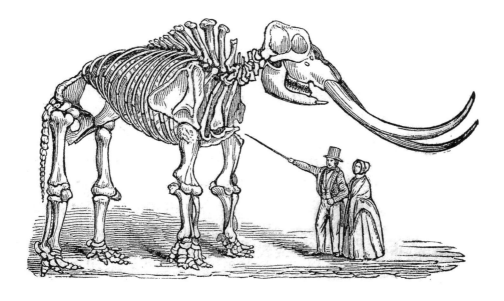

EXPLANATION
Terrestrial Contours
From Sea level to 1,000 feet   White
.   1,000 feet to 2,500 .
.   2,500 .   .   5,000 .
.   5,000 .   10,000 .
.   10,000 .   .20,000 .
Above 20,000 feet
The Marine Contour of 1,000 feet
is shown by a dotted line
Pasture lands shown thus
Forest
Desert
The boundaries and reference numbers
of the Sub-regions are shewn in Red.

London ; Macmillan & Co.

Alfred Russel Wallace (1823–1913), *The Geographical Distribution of Animals,* **1876; The neotropical region.**

some uniformitarians who claimed the Earth to be *infinitely* old. Finally, throughout this entire period embryologists and palaeontologists, particularly in France and the German-speaking countries, developed wide-ranging schemes by which the fossil history of animal species, the narrative of animals' embryonic development, and the new science of evolution might be unified into an all-encompassing scheme of animal life.

It was in the context of this intellectual ferment that two British naturalists changed biology forever. Charles Darwin was born in 1809, and although it was initially intended that he train as a doctor, his interests soon veered across into natural history. He made his scientific name as the zoologist on the 1831–1836 voyage of HMS *Beagle* during which he reported on marine fossils left high and dry in mountain rock strata, the bones of extinct fauna found alongside modern-looking mollusc shells, the apparently stable nature of the Earth, a single origin of the human species, and most famously the subtle variations between the finches of the Galapagos Islands. He probably started to formulate his theory of natural selection by 1838, and described it in an unpublished paper in 1844.

Born in 1823, Alfred Russel Wallace (see page 148) had travelled extensively in Brazil (1848–1852) and what is now Malaysia and Indonesia (1854–1862), documenting and collecting thousands of specimens. It was during the course of these explorations that he began to develop his own theories about natural selection. In 1855 he wrote in tree-like terms of 'branching lines of affinity, as intricate as the twigs of a gnarled oak', and began to describe genealogical trees of animal species, and the principles behind drawing them. However, Wallace was as much a geographer as a zoologist, and he discussed animals' environments, ecology and distribution as enthusiastically as their genealogical interrelationships. Indeed, he was the key instigator of the modern field of biogeography.

Darwin and Wallace published together on their new theory in 1858, and encouraged by what he had read in earlier works by Wallace, Darwin published *On the Origin of Species* on 24 November 1859. Throughout the

gestation of the theory of natural selection the two men had encouraged each other along the way, and Darwin certainly emphasized how similar their theories had been. They were to remain lifelong correspondents and mutual supporters, as their big idea began to reshape the scientific world.

Darwin's notebooks and the *Origin* contain surprisingly modern-looking schematic trees of animal relationships, although admittedly not many of them. The *Origin*, for example, contains only one diagram of any kind (see pages 132–133) – a tree of sorts. He seemed to think the tree motif was already sufficiently established for it to become the preferred form of representation of animals' relatedness, although he himself suggested that coral might be a better analogy, because coral polyps (like living species) grow atop a substrate of *dead* ancestors, not a living stem. In fact, perhaps the first published modern evolutionary tree had appeared the year before the *Origin*, hidden away in a dark corner of Heinrich Georg Bronn's *Entwickelungs-Gesetze der Organischen Welt* (see page 131). In some ways this Teutonic prefiguring of the evolutionary tree was appropriate because, as will become clear, it was another German, Ernst Haeckel (see page 162), who was to become both Darwin's most obsessive supporter, and the most avid creator of evolutionary trees in history.

Albert Günther (1830–1914), *Report on the Deep-Sea Fishes Collected by H.M.S.* **Challenger** *during the Years 1873–1876*, 1887; *Neobythites grandis* (1,875 fathoms) and *N. ocellatus* (350 fathoms).

PLATE I.

Fig. 1.

Fig. 5.

Fig. 2.

Fig. 3.

Fig. 4.

Fig 6.

M. F del.*   Published by A. Constable & C° Edin.*   W.H. Lizars Sc.

Left: John Fleming (1785–1857),
***The Philosophy of Zoology,* 1822;**
Bats and sea mammals.

John Fleming was a church minister
as well as a zoologist, and was
never able to reconcile his scientific
discoveries with his religious beliefs.
Certainly the fact that bats and
cetaceans are related mammals strains
the credulity of even the most ardent
supporter of evolution.

**Opposite: Friedrich Justin
Bertuch (1747–1822),** *Bilderbuch
für Kinder,* **1805; Squid and
octopuses.**

The title of Bertuch's weighty
encyclopedia-by-instalments,
'Picture book for children containing
a pleasant collection of animals,
plants, flowers, fruits, minerals...
appropriate to the intelligence of a
child' belies the quality and beauty
of its images.

Fig. 1.

Fig. 2.

Fig. 5.

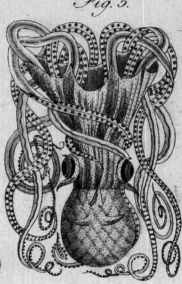

Fig. 3.

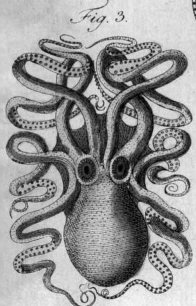

Fig. 4.

J. J. Schmuzer.

Carl Eduard von Eichwald (1795–1876), *Zoologia Specialis*, **1829; Tree of animal life.**

Born in present-day Latvia, Eichwald is often credited with creating the first true evolutionary tree, drawing on the written descriptions of Peter Simon Pallas (see page 57). Indeed, he did write of animals as changing over time, and having their origin in simple marine forms, although his exact meanings are sometimes obscure. The reticence of early eighteenth-century biologists to fully accept that animal groups divide over time is certainly evident in Eichwald's tree – mainly trunk, with small stubby branches. Thus are the affinities of different species emphasized above their divergence in this weatherbeaten bole.

Henry Thomas De la Beche (1796–1855), *Duria Antiquior* **(lithograph), 1830.**

The first, and still one of the most charming, artistic representations of a prehistoric scene, 'A More Ancient Dorset' draws on the fossil discoveries and reconstructions of Mary Anning in southwest England. The image is remarkable, not only because it accurately captures the scientific knowledge of the time, but also for its depiction of ancient creatures actually interacting with each other. The largest inhabitant of this part of ancient Dorset is a ravenous ichthyosaur, with a hint of glee on his face as he summarily dispatches a plesiosaur.

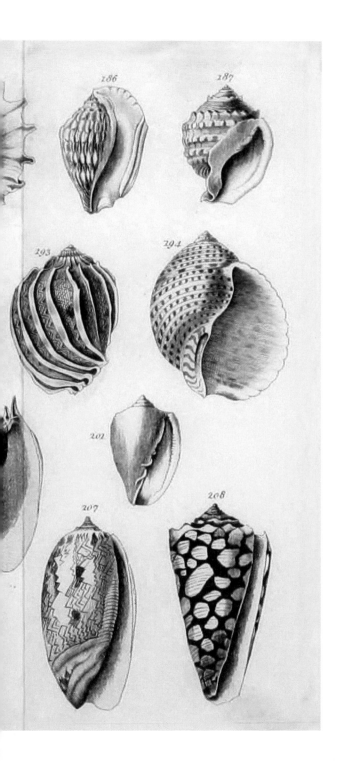

Jean-Baptiste Lamarck (1744–1829) and Anna Atkins (1799–1871), *Genera of Shells* **(English translation), 1833.**

Although Jean-Baptiste Lamark (see page 86), the author of this work, is most famous for his contribution to evolutionary theory, his zoological career started and ended with studies of invertebrate diversity. The English translation of this late work was beautifully illustrated by the British botanist, illustrator and later photographer, Anna Atkins, while still in her early twenties (Atkins may be the first woman ever to have taken a photograph.)

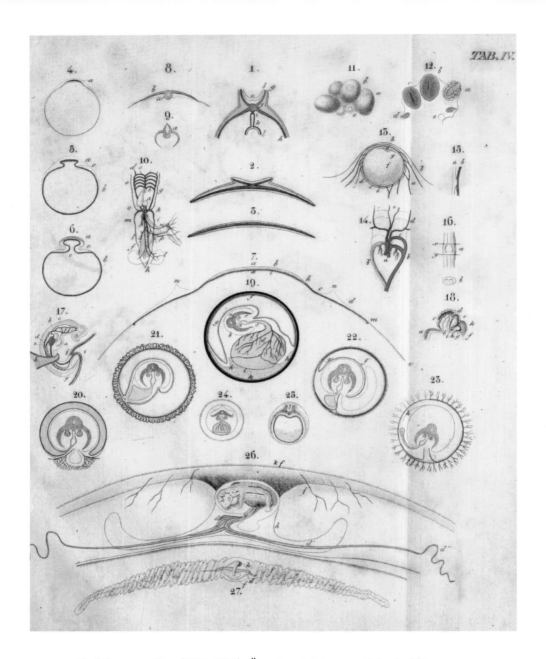

Karl Ernst von Baer (1792–1876), *Über Entwickelungsgeschichte der Thiere. Beobachtung und Reflexion,* **1838; Symmetries in the growth of vertebrates.**

The eighteenth century saw great advances in descriptive embryology – not only could species be studied changing over evolutionary time, but individual animals could also be studied developing *in utero* or *in ovo*. The Estonian embryologist Karl Ernst von Baer was seminal in realizing that these processes of species evolution and embryonic development must be linked, an interdisciplinary field of study still extremely active today.

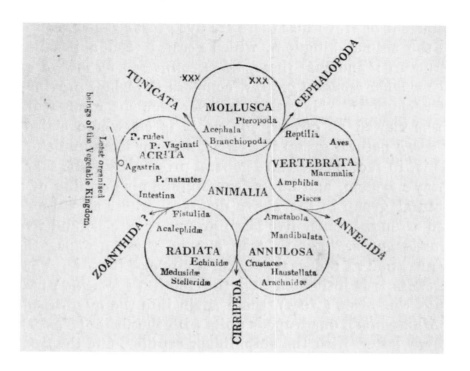

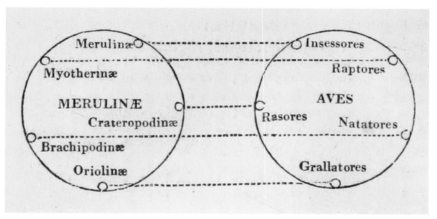

William Swainson (1789–1855), *A Treatise on the Geography and Classification of Animals,* **1835.**

Swainson was a proponent of the strange 'Quinarian System' of animal classification, by which all animal groups were claimed to be divisible into five subgroups, no more, no fewer. These were represented graphically by interacting circles, whose arrangement corresponded to their members' affinities and supposed advancement. An entirely unsubstantiated form of zoological numerology, it is not clear why the Quinarian System became even briefly popular.

TIMELINE

1793: Edward Hitchcock born

1823: *Geology of the Connecticut Valley* published

1825: Leaves the ministry

1840: *Elementary Geology* (first edition) published

1864: Edward Hitchcock dies

Edward Hitchcock

The Crown of Creation

Edward Hitchcock was the personification of the struggle between science and Christianity in the nineteenth century. After a stint as a devout Congregationalist pastor, he embarked on a successful career in which he was to face, head-on, the alarming inconsistencies between science and theology. In 1825, aged thirty-two, he became professor of Chemistry and Natural History at Amhurst College in Massachusetts, and soon instigated a variety of geological and palaeontological projects, as well as state-wide surveys in the earth sciences.

It rapidly became clear to Hitchcock that the chronology of the Bible fundamentally disagreed with the rocky evidence before his own eyes. In an attempt to reconcile the apparent antiquity of the Earth with the several-thousand-year-old world described in Genesis, he developed his own rather contrived reinterpretation of Bible chronology. He accepted that the six-day creation was a fallacy, and that Noah's flood was not the mechanism behind rock stratification and fossil deposition (before this time, fossils were often claimed to be antediluvian monsters eradicated by the Creator's deliberate inundation). Also, using a minor linguistic technicality, Hitchcock claimed that the biblical 'days' of creation signified six protracted periods of what we would now call 'geological time'.

Yet Hitchcock's god remained the creator and driver of organismal change, and although he depicted animal change and extinction over time in his strikingly tree-like 'Paleontological Chart' of 1840 (see opposite), he denied that such change could have solely mundane causes. Also, he was certain man does not share ancestry with the beasts.

In the later years of Hitchcock's life, Darwin and Wallace's ideas were to make his theological-scientific balancing act collapse. The world is indeed old, but it is the uncaring hand of natural selection which moulds change in those who inhabit it, not the hand of God. After the publication of *On The Origin of Species*, none of his books ever again contained his elegant and influential 'Paleontological Chart'.

Edward Hitchcock (1793–1864), *Elementary Geology*, 1840; Trilobites.

Fig. 54.

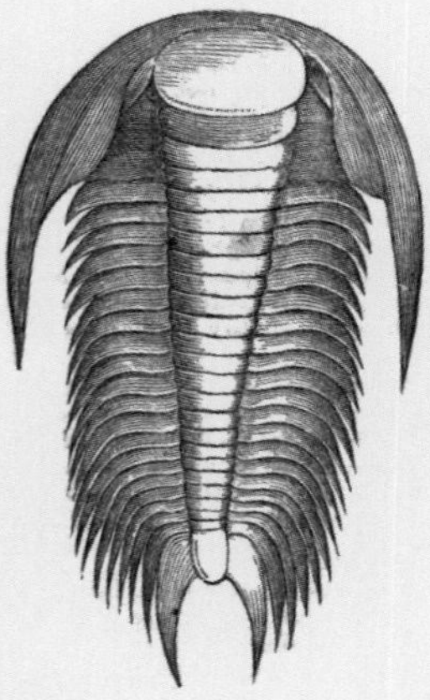

Fig. 55.

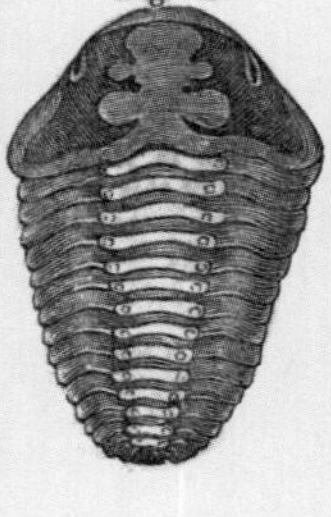

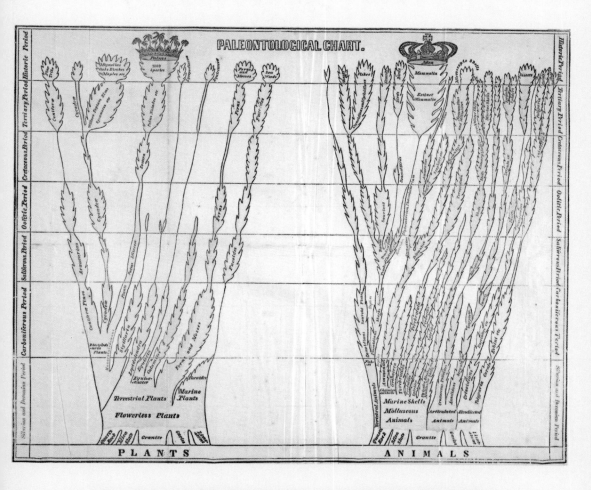

Edward Hitchcock (1793–1864), *Elementary Geology,* **1840; Paleontological Chart.**

Published in most editions of Hitchcock's much-reprinted *Elementary Geology*, his
'Paleontological Chart' appears as close as one can come to an evolutionary tree without
accepting non-deistic origins and processes for life. Indeed, he did refer to the chart as
'a tree'. It depicts the radiation of the families of plants (on the left) and animals (on the
right) upwards through the span of geological time. Some of the eras are familiar to us
today – the Carboniferous and Cretaceous, for example – while others, such as the 'Saliferous'
and 'Oolitic', sound archaic and mysterious. After the initial God-made radiation at the root
of each evolutionary 'bush', subsequent branching is allowed only when the evolutionary
relationships between groups could not reasonably be denied. Unsurprisingly, the (literal)
crown of animal creation is assumed to be the Mammalia, but in a strangely arbitrary attempt
to give the chart symmetry, it is the palms that are deemed to be monarchs of the plant world.

American Flamingo

John James Audubon (1785–1851), *The Birds of America,* **1827–1838;
Flamingo (above); Columbia jay (opposite).**

Every single image from Audubon's titanic – life-sized, indeed – *Birds of America*
(see page 103) is striking and beautiful. Here, the flamingo and Columbia jay
provide a pleasing chromatic juxtaposition.

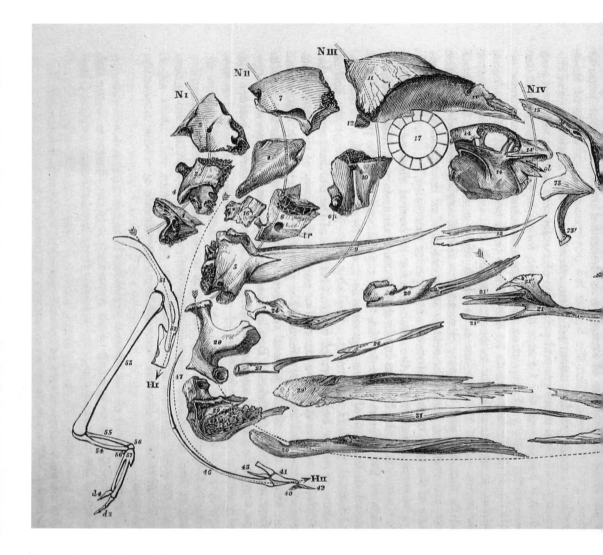

Richard Owen (1804–1892), *On the Archetype and Homologies of the Vertebrate Skeleton*, **1848; Homologies of the skeleton of an ostrich.**

Richard Owen was one of the most important figures in nineteenth-century vertebrate morphology, and dedicated great effort to seeking the organizational scheme underlying the structure of backboned animals. All vertebrates are constructed from the same set of anatomical precursors, and Owen was fascinated by how these precursors could generate creatures as diverse as fish, turtles, birds and whales. Owen sought to extrapolate back from modern animal structure to a single common ancestor of all vertebrates with an ordered, regular body structure from which all its descendants' body plans are derived. We now think that, while vertebrates do indeed share a single, long-lost ancestral species, its body plan was probably not as satisfyingly regular and neatly organized as Owen had hoped.

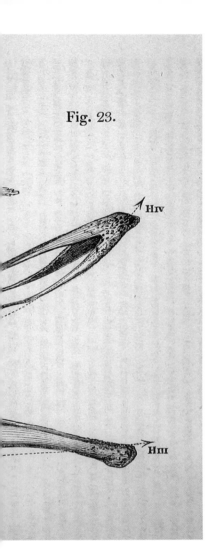

Fig. 23.

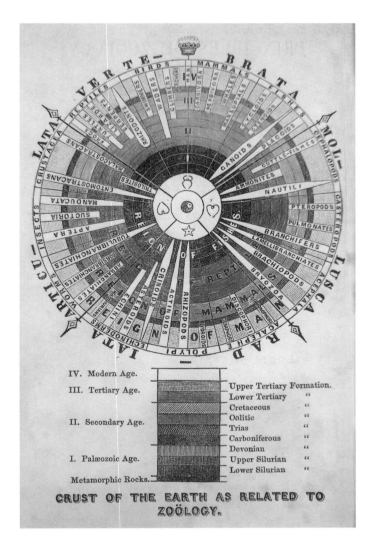

Louis Agassiz (1807–1873) and Augustus Gould (1805–1866), *Principles of Zoölogy*, **1851; Crust of the Earth as related to Zoölogy.**

Like William Smith (see page 90) and George Lyell, the Swiss scientist Louis Agassiz (see page 102) studied how different rock strata contain, and may be distinguished by, animal fossils from particular epochs. However, this diagram, almost zodiacal in its regularity, oversimplifies the messy pattern of animal evolution in its attempt to make its point.

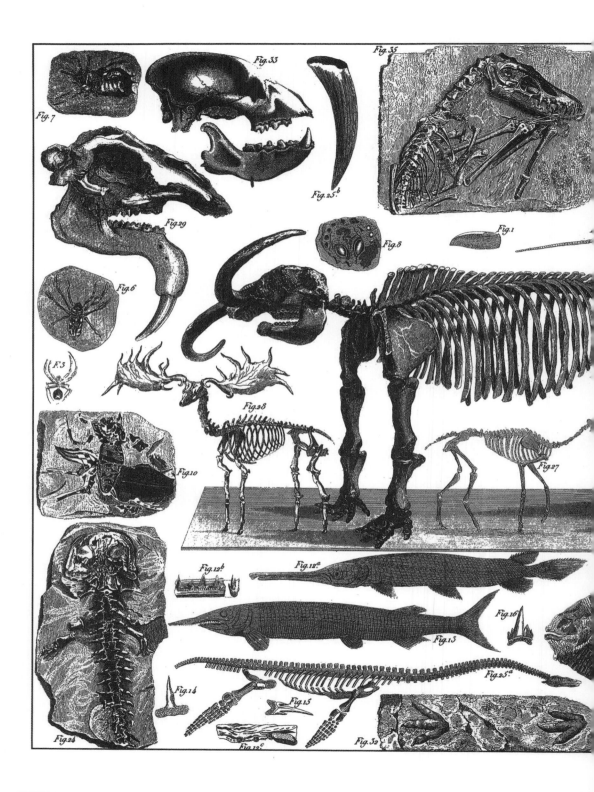

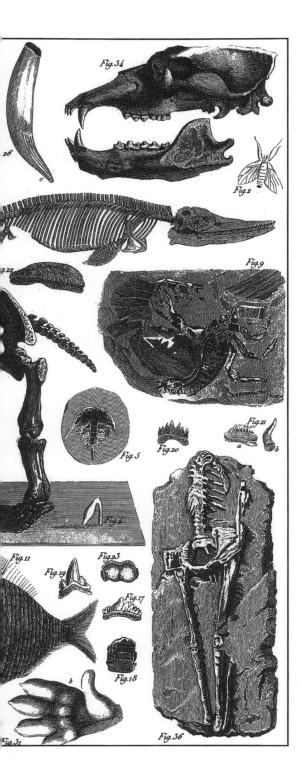

Anonymous, *Brockhaus and Efron Encyclopedic Dictionary*, **1890–1907; Fossil vertebrates.**

Once a striking zoological illustration is created, it may have a life beyond its initial context. Many of the images in this artist's resource are near-facsimiles of previously published designs, and many have been published repeatedly up to the present day.

Aristide-Michel Perrot (1793–1879), *Tableau des Animaux et des Végétaux Éxistans Avant le Deluge* **(coloured print), 1844.**

No doubt inspired by De la Beche's *Duria Antiquior* (see page 109), Perrot's *Tableau* veers towards the scientifically suspect in places. Elephants coexist with plesiosaurs, bats with pterosaurs, and an ammonite shell perches vertically on the crowded beach.

Johann Georg Heck (1823–1887), *Iconographic Encyclopedia Of Science, Literature and Art*, 1851, vol. 1, p. 74; The Order of Life: 144 images showing the progression and systematization of life.

Below: Benjamin Waterhouse Hawkins (1807–1894), *The Crystal Palace Dinosaurs*, 1853–1855; Author's photograph.

One of the most unusual urban sights, 'The Crystal Palace Dinosaurs', as they came to be known, were constructed near the site in south London where the 'Crystal Palace' centrepiece of the 1851 Great Exhibition had been relocated. A scientific illustrator, Hawkins sought the advice of eminent palaeontologists before commencing these first full-size recreations of ancient animals, yet much was left to his own interpretation – the texture of their skin, for example. Now often noted for its scientific inaccuracies, this much-loved diorama is really a testament to its creator's success in representing the scientific knowledge of his time.

Josiah Nott

Polygenism, Race and Slavery

For years before Darwin and Wallace's discovery of natural selection, scientists had pondered the origin of humankind. Most proponents of evolution believed, as biologists do today, that modern humans have a single ancestral origin, and thus are all related. Indeed, the fact that members of all ethnic groups can inter-breed to produce healthy, fertile children meant that, according to prevailing ideas, all humans are a single species. Yet for a brief period in the mid-nineteenth century, one person almost derailed this prevailing 'monogenic' theory of human origins.

A practising physician-surgeon in Mobile, Alabama, selective refuter of Christianity, and supporter of slavery, Josiah Nott dominated what was called the American School of Ethnology. Prevailing social attitudes meant questions of race and superiority were debated in a more heated atmosphere in the United States than in Europe. Slavery was ever-present and many sought justifications for its continuation.

Nott rejected evolution, and instead reverted to an almost bestiary-like form of animal classification, which he could then stretch to support his theories. For example, he was the most assertive proponent of 'polygenism', the idea that each of several 'human races' was a distinct entity separately created by God.

He disregarded scientific evidence and scriptural narrative with equal abandon, stating that mixed-race people were akin to mules, and that Genesis recounted only the history of a small subset of mankind descended from a white Caucasian Adam. As a result, Nott claimed the evidence showed the white man was justified in dominating the black man, whose attributes render him the perfect slave.

Despite Nott's fervour, polygenism was short-lived. He could not deny that Darwin's theories were the death-knell for a separate origin for human races, and ceased his research and writing. Soon, he could no longer tolerate living in what he called 'Negroland', and moved his practice to the relatively white and bible-less New York City.

However, polygenism, and the racial theories it supported, did not disappear entirely.

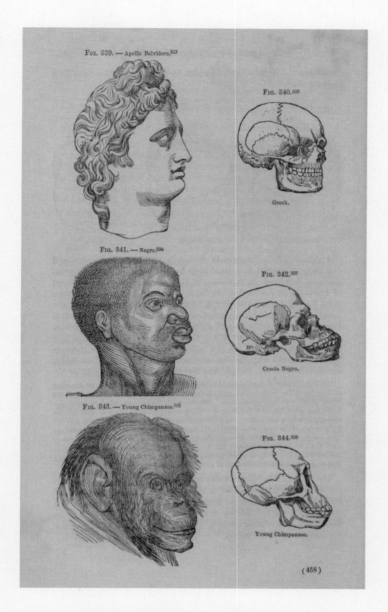

FIG. 339. — Apollo Belvidere.[563]

FIG. 340.[565]

Greek.

FIG. 341. — Negro.[564]

FIG. 342.[567]

Creole Negro.

FIG. 343. — Young Chimpanzee.[565]

FIG. 344.[568]

Young Chimpanzee.

(458)

Josiah Nott (1804–1873), George Gliddon (1809–1857) and others,
Types of Mankind, **1854; Apollo Belvedere, negro and chimpanzee.**

Shocking today, the implication of this image is clear. In the decades when the
mechanisms underlying the process of evolution remained unexplained, the
large amounts of newly acquired biological information could be used to justify
almost any conclusion. Here, the anatomy of a chimpanzee and black and
Caucasian humans are unquestioningly compared, without reference to any
evidence regarding their origins, adaptations or ecology.

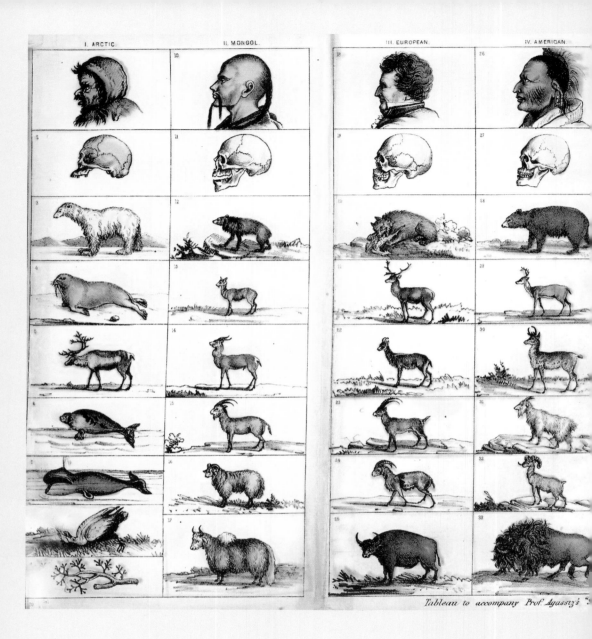

I. ARCTIC.   II. MONGOL.   III. EUROPEAN.   IV. AMERICAN.

Tableau to accompany Prof. Agassiz's

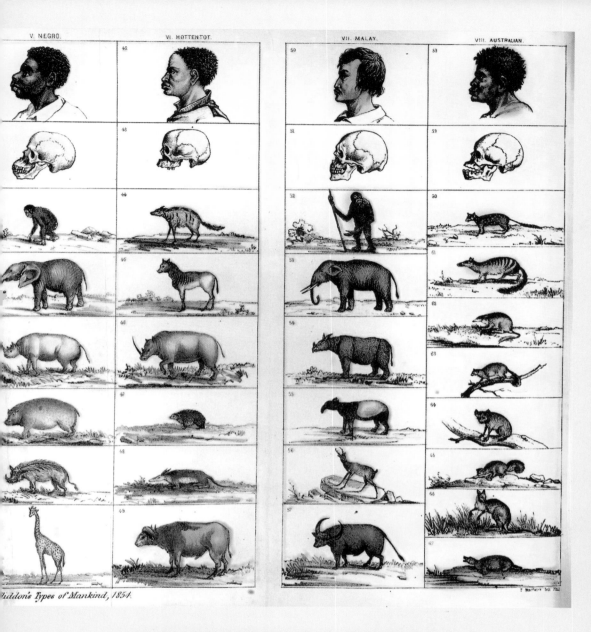

Gliddon's Types of Mankind, 1854.

Josiah Nott (1804–1873), George Gliddon (1809–1857) and others,
Types of Mankind, **1854; 'Tableau to accompany Dr Agassiz's sketch'.**

Animal classification in the support of social dogma. This chart accompanied
an article by Louis Agassiz (see page 102) in Nott and Gliddon's 1854 review
of the ethnography, geography and origins of the human races. Chiming with
Nott's polygenic beliefs, each of eight columns presents the animal inhabitants
of one of the world's major geographical regions. Every region is depicted as
being inhabited by a distinct, discrete type of human, just as it is home to its
own species of carnivorous and grazing animals.

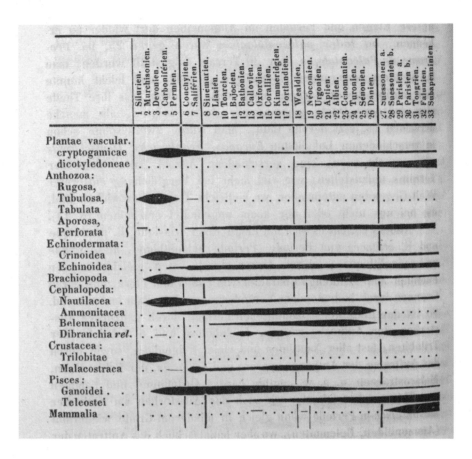

Heinrich Georg Bronn (1800–1862), *Untersuchungen über die Entwickelungs-Gesetze der Organischen Welt wahrend der Bildungs-Zeit unserer Erd-Oberfläche,* **1858 (above and opposite).**

Published the year before *On the Origin of Species* (see page 104), and known to be in Darwin's own collection, Bronn's lengthy book was a wide-reaching survey of existing geological and zoological knowledge. It featured varied representations of past changes in the diversity of animal groups over time, including the 'spindle diagram' on this page. However, most important is the tiny tree tucked away in a corner of one of its later pages (see opposite). Unlike the tree produced by Eduard von Eichwald (see page 108), this image emphasizes the progressive diversifications of animal forms, with branches and sub-branches appearing at apparently irregular intervals. Bronn's reclusive shrub can thus be said to be the first truly 'modern' evolutionary tree.

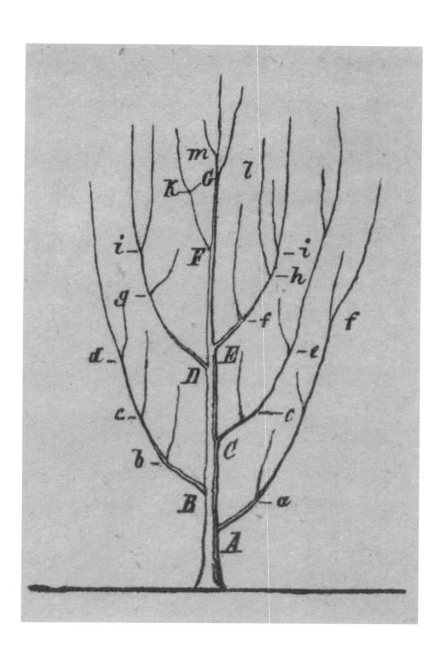

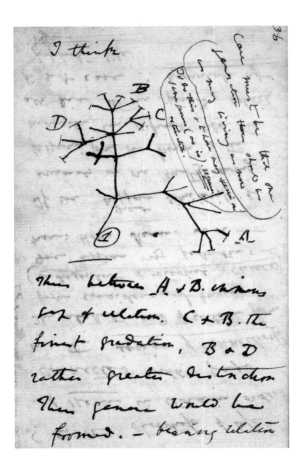

Charles Darwin (1809–1882), *Notebook 'B' excerpt,*
c.1837–1838 (above); *On the Origin of Species,* **1859;**
Sole figure (right).

Although Darwin had long considered the possibility that
the evolutionary process could be represented in tree-like form,
he himself drew few trees – and indeed doubted that sufficient
fossil evidence would ever be available for credible trees of actual
animal groups to be created. The image above shows the now-
famous 'I think' page of his notebooks from as far back as the
1830s, in which he uses a tree-form to illustrate how the pattern
of evolutionary change and diversification creates species with
differing degrees of relatedness. In contrast, the only figure in *On
the Origin of Species* (see opposite) is an austere thing, certainly
lacking the fluidity of the little tree (see page 131) of Heinrich
Bronn from the previous year. Darwin mainly uses his diagram
to explain divergence of sub-populations *within* a species, but
later explains how the tree may be traced back to larger twigs
and branches, eventually encompassing all living things.

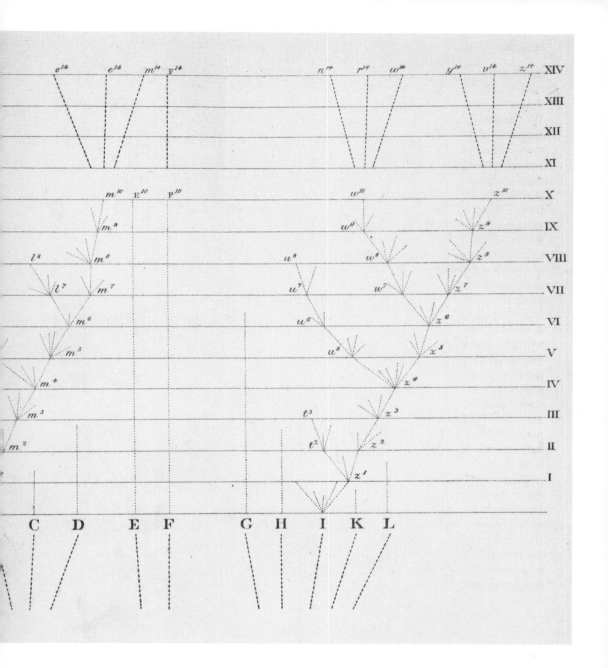

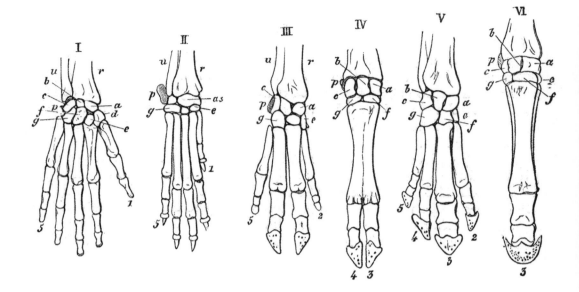

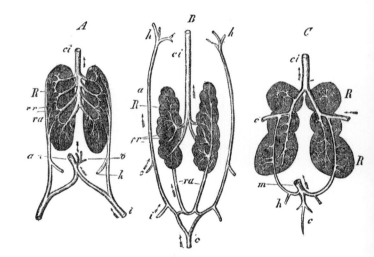

Karl Gegenbaur (1826–1903), *Grundzüge der Vergleichenden Anatomie*, 1859; Skeleton of the manus (forefoot) of various mammals (above); Posterior portion of the venous system of the frog, alligator and bird (right); Side view of the skulls of an ostrich, a crocodile and a python (opposite above); Differences in the arrangement of the ribs and transverse processes in the Teleostei [bony fish] (opposite below).

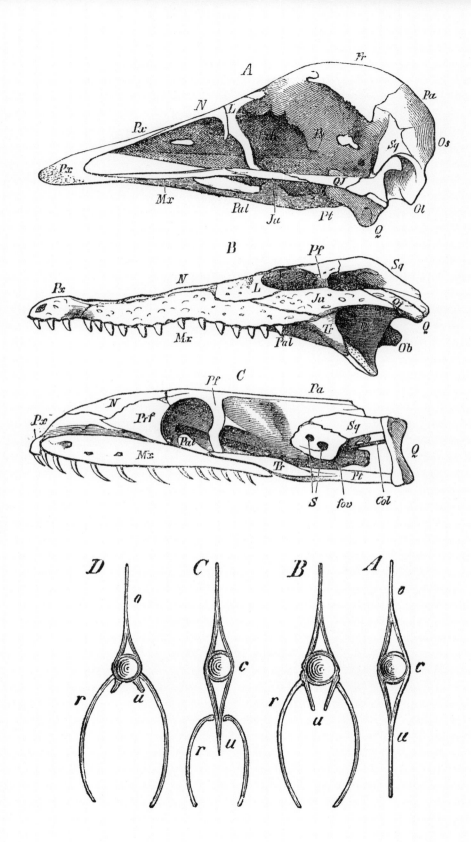

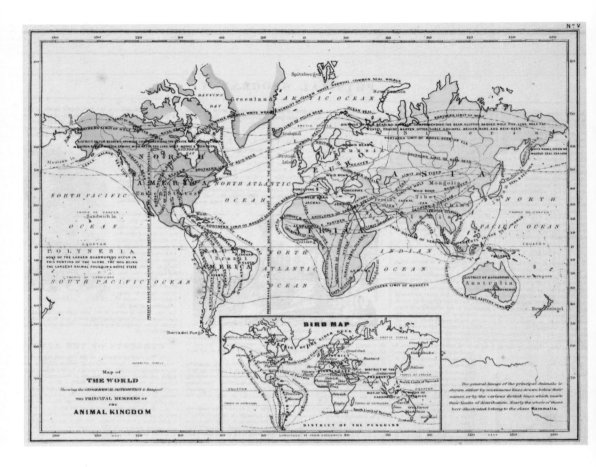

Alvin Jewett Johnson (1827–1884), *Johnson's New Illustrated Family Atlas,* **1860; Map of the World Showing the Geographical Distribution and Range of the Principal Members of the Animal Kingdom.**

It was to be Alfred Russel Wallace (see pages 104 and 148) who developed our modern view of organisms in the context of particular geographical regions, but this earlier image shows this was not an entirely new field of inquiry.

Philip Henry Gosse (1810–1888), *A History of the British Sea-Anemones and Corals,* **1860; Anemones.**

Having worked in fisheries at home and abroad, the British marine biologist Philip Gosse was the inventor of the marine aquarium, where his subjects could be observed, and drawn, at leisure.

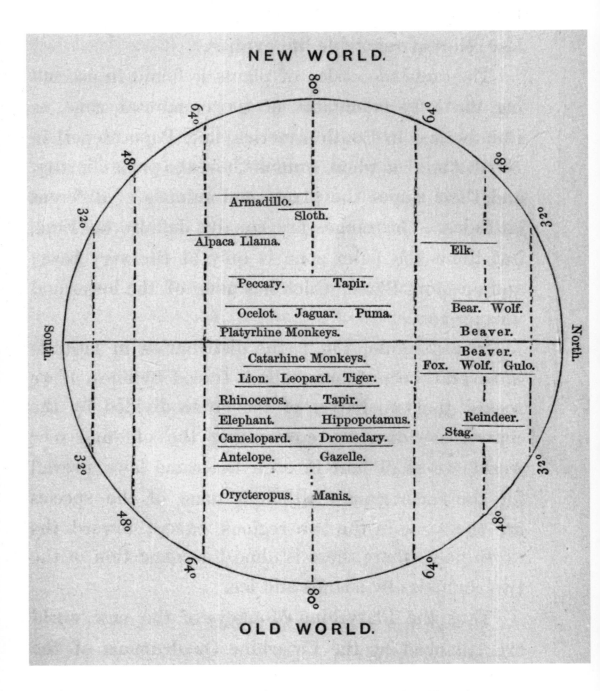

NEW WORLD.

OLD WORLD.

John Phillips (1800–1874), *Life on the Earth: Its Origin and Succession*, 1860; Distributions in latitude of several races of Mammalia.

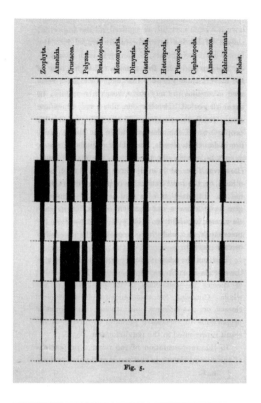

Fig. 5.

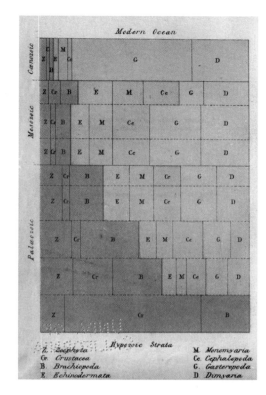

Modern Ocean

Z Zoophyta
Cr Crustacea
B Brachiopoda
E Echinodermata

M Monomyaria
Ce Cephalopoda
G Gasteropoda
D Dimyaria

Hypozoic Strata

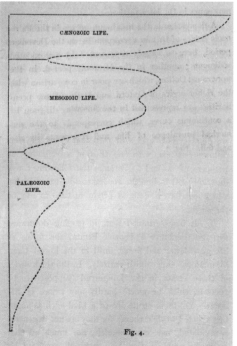

CÆNOZOIC LIFE.

MESOZOIC LIFE.

PALÆOZOIC LIFE.

Fig. 4.

John Phillips (1800–1874), *Life on the Earth: Its Origin and Succession*, 1860; A scheme of proportionate life for the Lower Paleozoic Strata (above left); Successive systems of marine invertebral life (above right). A continuous curve that corresponds to the numerical prevalence of life (left).

Phillips was an English geologist who meticulously compiled data from around the world to create the first global system of standardized geological epocha. To do this, he developed many novel forms of organization of extant and extinct animal species. On these pages he surveys the distribution of living species by latitude and the prevalence of fossils in ancient strata, and on subsequent pages considers the vagaries of the rise and fall of different animal groups over time.

John Gould (1804–1881), *The Mammals of Australia,* **1845–1863;**
Ornithorhynchus anatinus.

Over the years, the charming platypus has presented more challenges to taxonomists
than any other species. An assemblage of superficially unrelated characteristics, it is now
known to be an extremely specialized mammal that diverged from the main mammalian
'stem' very early. It suckles its young, which defines it as a mammal, but lays eggs, possibly
reflecting the retention of an ancestral mammalian characteristic. Its webbed feet and
flat tail have evolved convergently with waterfowl and beavers, respectively, but its bill is a
unique electroreceptive organ used to seek its aquatic invertebrate prey. The platypus even
has the distinction of being one of very few venomous mammals.

John Gould (1804–1881), *The Mammals of Australia,* **1845–1863;**
Thylacinus cynocephalus.

The thylacine or marsupial wolf is one of the most dramatic known examples of
convergent evolution – the evolution of similar features by distantly related species
that nonetheless have similar ecological niches. The body shape, skeleton and, especially,
the skull of the thylacine are remarkably similar to those of dogs and wolves, despite its
genealogical affinities with kangaroos and wombats. The story of this unique animal
is, however, one of the saddest in all zoology. Aggressively hunted by white settlers, the
thylacine was probably driven to extinction in the early twentieth century – although
tantalizing reports still occur of survivors in remote Tasmanian forests.

A Moravian friar, Gregor Mendel seems an unlikely person to have instigated the modern genetic revolution. We will probably never know what drove the Abbot of St. Thomas' Abbey in Brno to conduct the experiments in plant hybridization that demonstrated that inheritance of biological characteristics involves physical mechanisms that follow arithmetic laws. Although his 1865 'Experiments in Plant Hybridization' was published only six years after *On the Origin of Species*, it was largely ignored for forty years. In consequence, scientists' ability to organize the animal world according to genetics was delayed by decades.

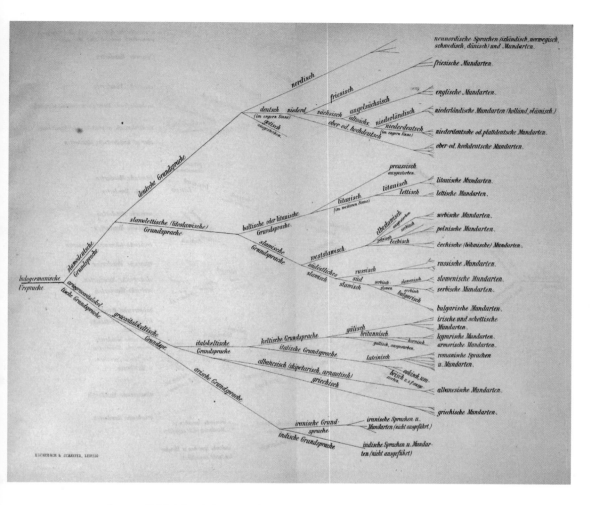

August Schleicher (1821–1868), *Die Darwinsche Theorie und die Sprachwissenschaft,* **1863;** *Stammbaum* **for Indo-European.**

Evolutionary biology has overlapped to a surprising extent with linguistics, and in this diagram the linguistician August Schleicher takes a Darwinian approach to the changes and diversification of the Indo-European languages of Europe and south and west Asia (*Stammbaum* means 'family tree'). Of course, as well as splitting, languages can share elements and possibly even merge, something that is now thought in some ways to be true of animal species, too (see pages 230–231). Also, it is possible that even the word 'evolution' was appropriated by eighteenth- century biologists from seventeenth-century studies of language change over time.

Imp. Becquet à Paris.

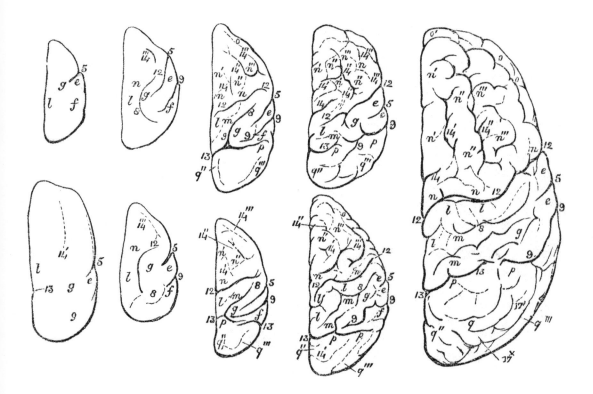

Above: Richard Owen (1804–1892), *On the Anatomy of Vertebrates,*
**1868; Upper views of the cerebral hemispheres of the human, a foetus
and other primates.**

A strange montage in which human cerebra are interspersed with those of other
primates. The top row, from left to right, depicts *Midas* (tamarin), *Callithrix*
(marmoset), *Macacus* (macaque) and a human infant, while the bottom row shows
the brains of a human foetus, a lemur, *Cebus* (capuchin monkey) and chimpanzee.
The adult human brain dominates the right side of the image.

Opposite: Alfred Frédol (1804–1863), *Le Monde de la Mer,*
1866; *Développement d'un oiseau.*

Frédol was the pseudonym of Alfred Moquin-Tandon, Professor
of Zoology and Director of the *Jardin des Plantes* in Paris.

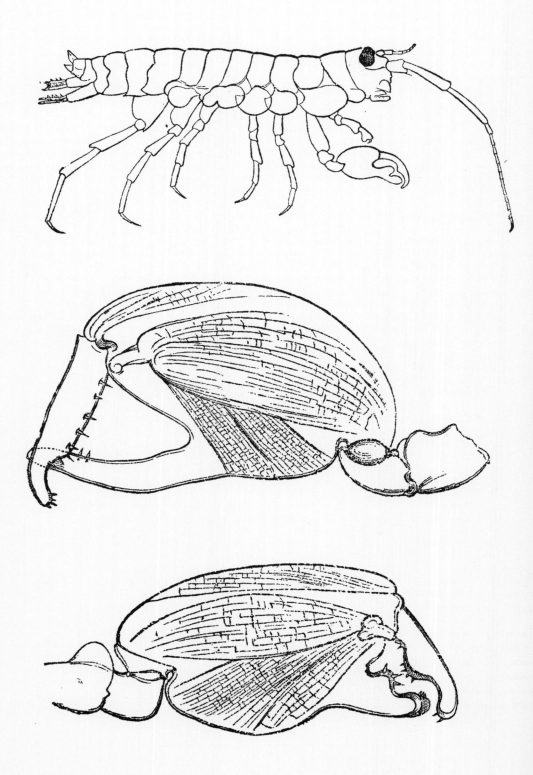

Above: St. George Jackson Mivart (1827–1900), *On The Genesis Of Species,* **1871; The walking leaf insect.**

Opposite: Charles Darwin (1809–1882), *The Descent of Man and Selection in Relation to Sex,* **1871;** *Orchestia darwinii* **showing the differently constructed chelae of the two male forms.**

Having established the theory of natural selection, Darwin extended his initial ideas considerably, and indeed established the conceptual framework on which much of modern evolutionary biology is based. He discussed the origin of life, the speed and pattern of evolution, the origins of humans, and sexual selection (a different mechanism of evolution distinct from natural selection). In this diagram, Darwin notes that evolution sometimes leads to males of different morphologies within the same species.

Alfred Russel Wallace

Animals in space, as well as time

Along with Darwin, the Welsh naturalist and explorer Alfred Russel Wallace was one of the two architects of the now-dominant theory of natural selection. However, he was also the scientist who gave zoological classification a geography, as well as a history.

Although he never enjoyed the financial security into which Darwin married, Wallace was able to finance a far more adventurous life. Inspired by stories of 'new lands', he participated in an expedition to uncharted regions of the Amazon in the late 1840s, as well as a career-defining, eight-year trip to what are now Malaysia and Indonesia.

One feverish night in 1858, Wallace suddenly realized how species evolve, and immediately wrote to his long-time correspondent, Charles Darwin, who he knew had been toying with similar concepts for years.

From that point onwards, it is difficult to separate the two men's roles in the theory of natural selection. They corresponded regularly, supported each other's ideas and published together in 1858, the year before the publication of *On the Origin of Species*. Wallace remained Darwin's greatest supporter, entirely content to be the less famous of the two.

As well as his pivotal yet understated role in the theory of natural selection, Wallace was a founder of the science of biogeography – the study of the geographical distribution of animals past and present. Drawing on the work of the British zoologist Philip Sclater, Wallace identified the six global regions of faunal distribution still in use today (see pages 150–151).

Indeed, the Oriental and Australian regions remain separated by the eponymous invisible 'Wallace Line', which he himself discerned partitioning the animal life of Borneo and Bali in the west and that of Sulawesi and Lombok in the east.

Alfred Russel Wallace (1823–1913), *Darwinism,* **1889; Variation of** *Icterus baltimore.*

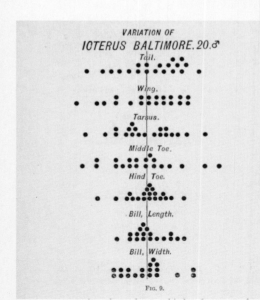

VARIATION OF
IOTERUS BALTIMORE. 20.♂

Tail.

Wing.

Tarsus.

Middle Toe.

Hind Toe.

Bill, Length.

Bill, Width.

FIG. 9.

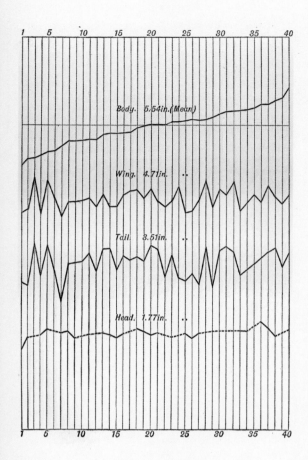

Alfred Russel Wallace (1823–1913),
Darwinism, 1889; Variation in body, wing,
tail and head size in forty male red-winged
blackbirds (left); Mimicry among insects
from the Philippines (below).

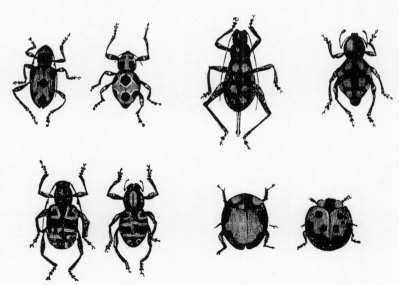

Alfred Russel Wallace (1823–1913), *The Geographical
Distribution of Animals*, 1876; a Brazilian forest, with
a range of native mammals.

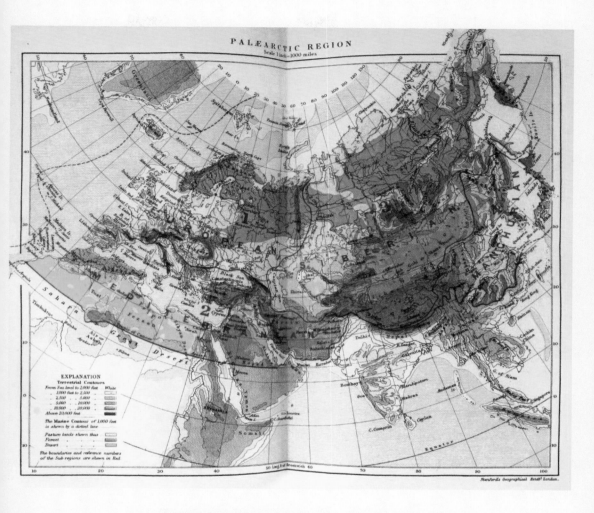

Alfred Russel Wallace (1823–1913), *The Geographical Distribution of Animals*, 1876; The Palaearctic region (above); A Malayan forest, with its characteristic birds (opposite).

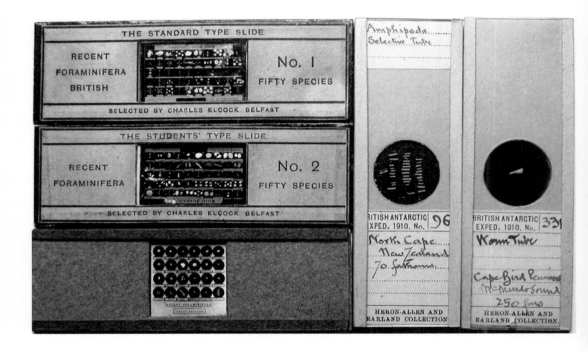

Harold Dalton (1829–1911); *Microscope mount including insect scales and single-celled animals,* **c.1875 (above); Micromosaic (opposite).**

Micromosaics enjoyed a brief flurry of popularity in the late nineteenth century. Insect scales, single-celled diatoms and other minutiae were painstakingly harvested, categorized and arranged on microscope slides, using needles and tiny air-blowers to create kaleidoscopic arrangements. Zoologists harvested zoological specimens from around the world to incorporate into these artistic mini-tableaux.

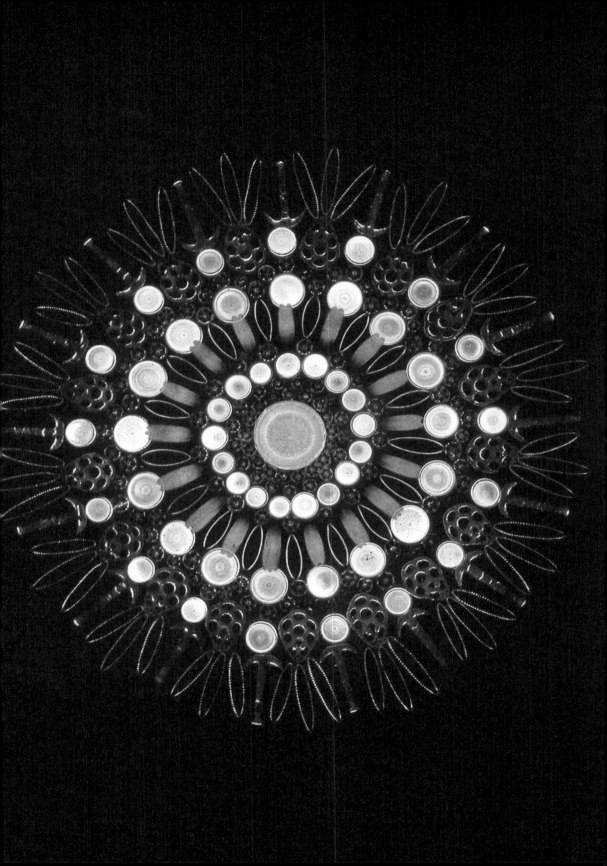

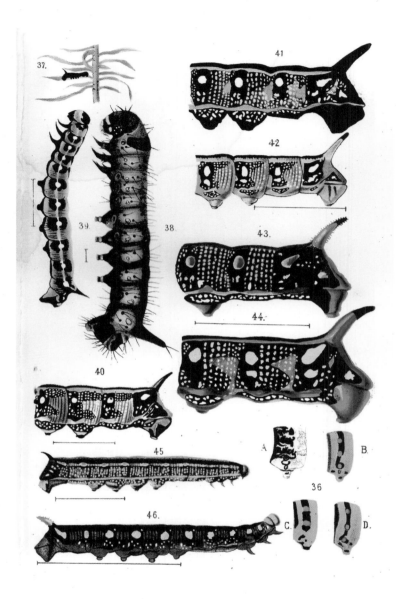

August Weismann (1834–1914), *Studies in the Theory of Descent*, **1882 (above and opposite).**

Weismann developed the 'germ plasm' theory of inheritance, by which offspring inherit traits solely by means of information carried in the germ cells – eggs and sperm. Today, this concept is the main refutation of the mode of evolution proposed by Jean-Baptiste Lamarck (see page 86) – in which influences on any part of an animal's body may be inherited by its offspring. Weismann's own interpretations of his germ plasm theory were, in fact, more subtle and complex.

	PRODUCTS OF EMOTIONAL DEVELOPMENT.		EMOTION.
50		50	
49		49	
48		48	
47		47	CIVILISED. SAVAGE. HUMAN.
46		46	
45		45	
44		44	
43		43	
42		42	
41		41	
40		40	
39		39	
38		38	
37		37	
36		36	PARTLY HUMAN.
35		35	
34		34	
33		33	
32		32	
31		31	
30		30	
29		29	SOCIAL.
28	Shame, Remorse, Deceitfulness, Ludicrous.	28	
27	Revenge, Rage.	27	
26	Grief, Hate, Cruelty, Benevolence.	26	
25	Emulation, Pride, Resentment, Æsthetic love of ornament, Terror.	25	
24	Sympathy.	24	
23		23	
22	Affection.	22	
21	Jealousy, Anger, Play.	21	
20	Parental affection, Social feelings, Sexual selection, Pugnacity, Industry, Curiosity.	20	PRESERVATION OF SPECIES OF S
19	Sexual emotions without sexual selection.	19	
18	Surprise, Fear.	18	
17		17	
16		16	
15		15	
14		14	
13		13	
12		12	
11		11	
10		10	
9		9	
8		8	
7		7	
6		6	
5		5	
4		4	
3		3	
2		2	E X
1		1	

Harrison

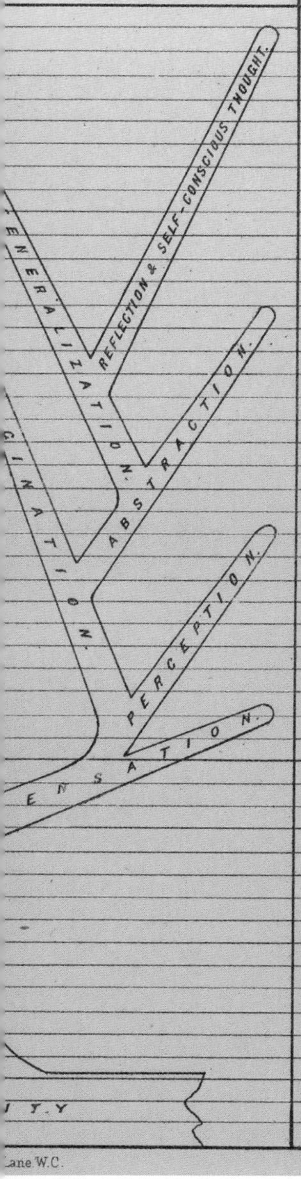

INTELLECT.		PRODUCTS OF INTELLECTUAL DEVELOPMENT.	THE PSYCHOLOGICAL SCALE.	PSYCOGENESIS OF MAN.	
	50				50
	49				49
	48				48
	47				47
	46				46
	45				45
	44				44
	43				43
	42				42
	41				41
	40				40
	39				39
	38				38
	37				37
	36				36
	35				35
	34				34
	33				33
	32				32
	31				31
	30				30
	29				29
	28	Indefinite morality.	Anthropoid Apes and Dog.	15 months.	28
	27	Use of tools.	Monkeys, a... Elephant	12 months.	27
	26	Understanding of mechanisms.	Carnivora, Rodents, and Ruminants	10 months.	26
	25	Recognition of Pictures, Understanding of words, Dreaming.	Birds.	8 months.	25
	24	Communication of ideas.	Hymenoptera.	5 months.	24
	23	Recognition of persons.	Reptiles and Cephalopods.	4 months.	23
	22	Reason.	Higher Crustacia.	14 weeks.	22
	21	Association by similarity.	Fish and Batrachia.	12 weeks.	21
	20	Recognition of offspring, Secondary instincts.	Insects and Spiders.	10 weeks.	20
	19	Association by contiguity.	Mollusca.	7 weeks.	19
	18	Primary instincts.	Larvæ of Insects, Annelida.	3 weeks.	18
	17	Memory.	Echinodermata.	1 week.	17
	16	Pleasures and pains.		Birth.	16
	15	Nervous adjustments.	Cœlenterata.		15
	14				14
	13				13
	12		Unknown animals,		12
	11	Partly nervous adjustments.	probably Cœlenterata.		11
	10		perhaps extinct.	Embryo.	10
	9				9
	8				8
	7	Non-nervous adjustments.	Unicellular organisms.		7
	6				6
	5				5
	4				4
	3	Protoplasmic movements.	Protoplasmic organisms.	Ovum and	3
	2				2
	1			Spermatozoa	1

(Left margin branching diagram labels: GENERALIZATION, REFLECTION & SELF-CONSCIOUS THOUGHT, ABSTRACTION, IMAGINATION, PERCEPTION, SENSATION. Centre annotation: ← Consciousness →)

Lane W.C.

George Romanes (1848–1894), *Mental Evolution in Animals,* **1883; Frontispiece.**

A friend and supporter of Darwin, Romanes was particularly interested in finding links between cognitive processes, and even consciousness, in animals and humans. At a time when the theory of natural selection was very new, and the brain remained poorly understood, this was a challenging topic in which to become embroiled.

Opposite: Ernst Haeckel (1834–1919), *Kunstformen der Natur*, **1904; Chiroptera – Fledetiere.**

Below: Ernst Haeckel (1834–1919), *Natürliche Schöpfungsgeschichte*, **1868;** *Die Familiengruppe der Katarrhinen.*

Ernst Haeckel

Trees of Species, Trees of Races

Professor of Zoology at Jena, Ernst Haeckel was a potent force in nineteenth-century zoology. An über-Darwinist, he took the theories of his more cautious English contemporary and created his own, more strident version – Darwinismus.

Originally an artist, and strongly influenced by the German Romanticists, Haeckel travelled the world collecting and painting the specimens that would support his theories of evolution and natural selection. He toured with a menagerie of skeletons and embryos, and his profusely illustrated books adorned the drawing-room tables of Europe.

For his time, Haeckel was unusually clear in his assertion that Man is just one species among many, and his zoological classifications are based on the idea that all life has evolved from a single, simple ancestor. Because of this, his evolutionary trees all have a single stem, and human inhabit their rarefied crowns.

Haeckel wrote extensively on the origins of life in the simplest creatures – his 'cell-souls and soul-cells' – and developed uncompromising views on the progressive nature of evolution, and the links between evolution and the burgeoning field of embryology.

To him, all progress towards perfection was the same, but he is now mainly remembered for over-extending his theorie into religion and race. When viewing the beauty of his zoological diagrams, it can be hard to believe this is the same man who described Judaism as being intermediate between 'primitive' paganism and 'advanced' Christianity, or wrote tha non-Europeans are '…physiologically nearer to the mammal – apes and dogs – than to the civilized European. We must, therefore, assign a totally different value to their lives.'

Although his writings were later lauded by the Nazis, Haeckel's art remains a glorious testament to the classificato traditions of nineteenth-century natural philosophers.

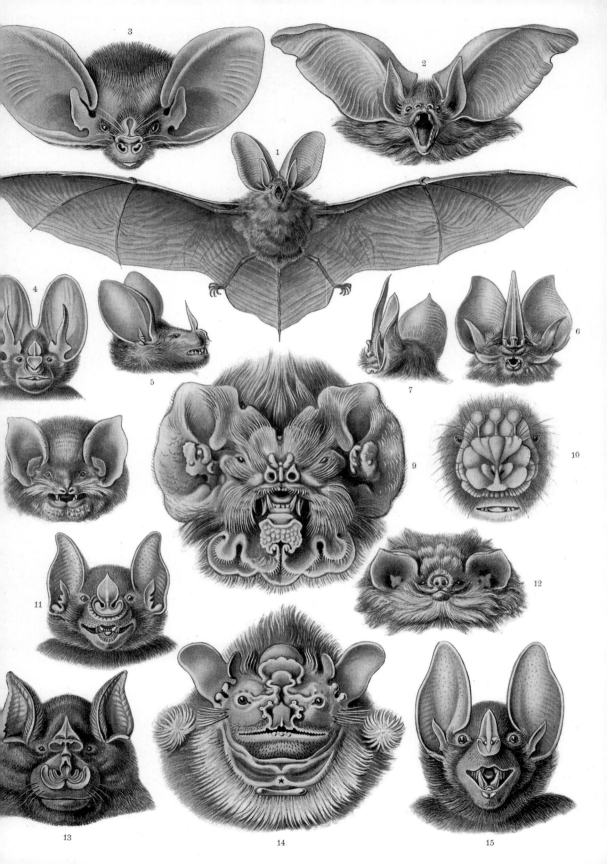

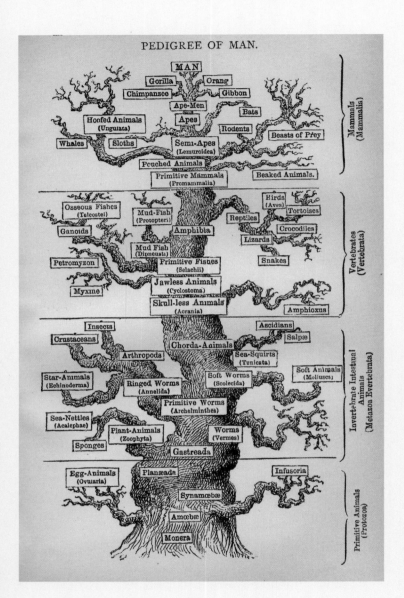

PEDIGREE OF MAN.

MAN

Gorilla · Orang
Chimpanzee · Gibbon
Ape-Men
Apes · Bats
Hoofed Animals (Ungulata) · Rodents
Whales · Sloths · Semi-Apes (Lemuroidea) · Beasts of Prey
Pouched Animals
Primitive Mammals (Promammalia) · Beaked Animals.

Mammals (Mammalia)

Osseous Fishes (Teleostei) · Mud-Fish (Protopteri) · Birds (Aves) · Tortoises
Ganoids · Reptiles
Amphibia · Crocodiles
Lizards
Mud Fish (Dipneusti) · Snakes
Petromyzon · Primitive Fishes (Selachii)
Jawless Animals (Cyclostoma)
Myxine · Amphioxus
Skull-less Animals (Acrania)

Vertebrates (Vertebrata)

Insects · Ascidians
Crustaceans · Salpæ
Chorda-Animals
Arthropods · Sea-Squirts (Tunicata)
Soft Worms (Scolecida) · Soft Animals (Molluscs)
Star-Animals (Echinoderma) · Ringed Worms (Annelida)
Sea-Nettles (Acalephæ) · Primitive Worms (Archelminthes)
Plant-Animals (Zoophyta) · Worms (Vermes)
Sponges · Gastreada

Invertebrate Intestinal Animals (Metazoa Evertebrata)

Egg-Animals (Ovularia) · Planæada · Infusoria
Synamœbæ
Amœbæ
Monera

Primitive Animals (Protozoa)

Left: Ernst Haeckel (1834–1919), *Anthropogenie oder Entwickelungsgeschichte des Menschen*, **1874;** *Stammbaum des Menschen.*

Few could devise evolutionary trees like Haeckel, and this single gnarled example – the 'Pedigree of Man' – summarizes the entirety of animal creation. In a way that would be unacceptable today, it also allocates animals to different 'grades' of advancement at different heights – single-celled animals, invertebrates, vertebrates, mammals – ascending towards the peak of creation which is, of course, humans.

Opposite: Ernst Haeckel (1834–1919), *Kunstformen der Natur*, **1904; Discomedusae.**

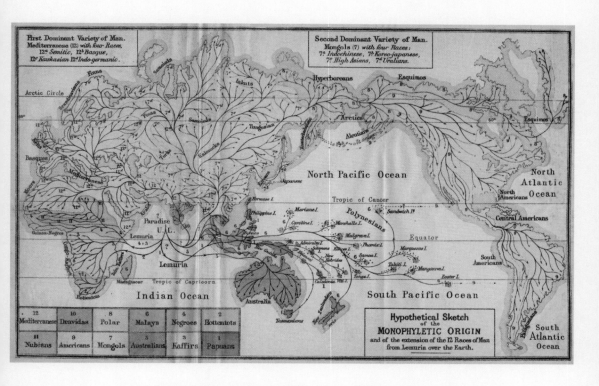

Above: Ernst Haeckel (1834–1919), *Natürliche Schöpfungsgeschichte,* **1868; Hypothetical sketch of the monophyletic origin and of the extension of the 12 Races of Man from Lemuria over the Earth.**

Opposite: Ernst Haeckel (1834–1919), *Kunstformen der Natur,* **1904; Ostraciontes – Kofferfische.**

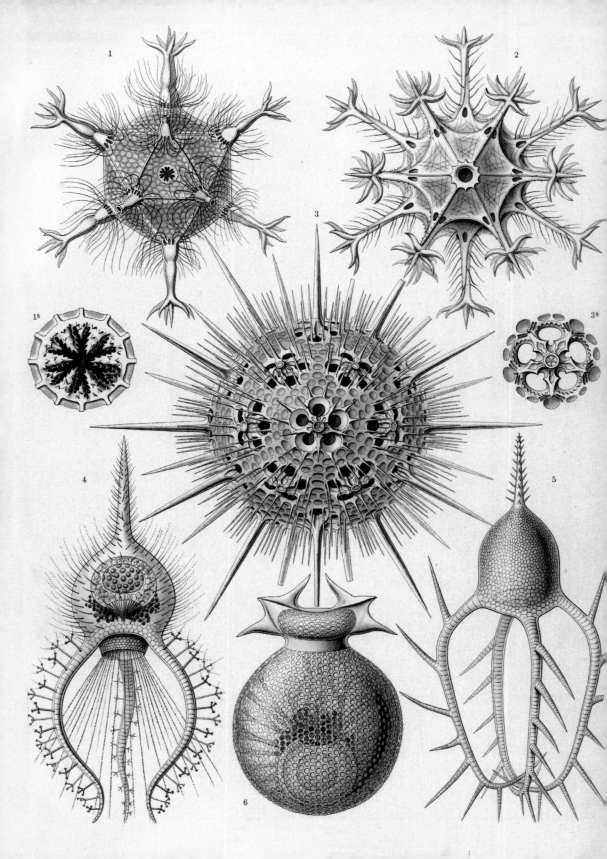

Right: Ernst Haeckel (1834–1919), *Generelle Morphologie der Organismen*, **1866.**

This evolutionary tree, unusually, classifies the mammals according to the structure of their placenta – their most anatomically diverse organ.

Opposite: Ernst Haeckel (1834–1919), *Die Radiolarien*, **1862.**

Radiolarians are single-celled organisms that secrete complex and distinctive silica skeletons, and were Haeckel's favourite artistic subjects.

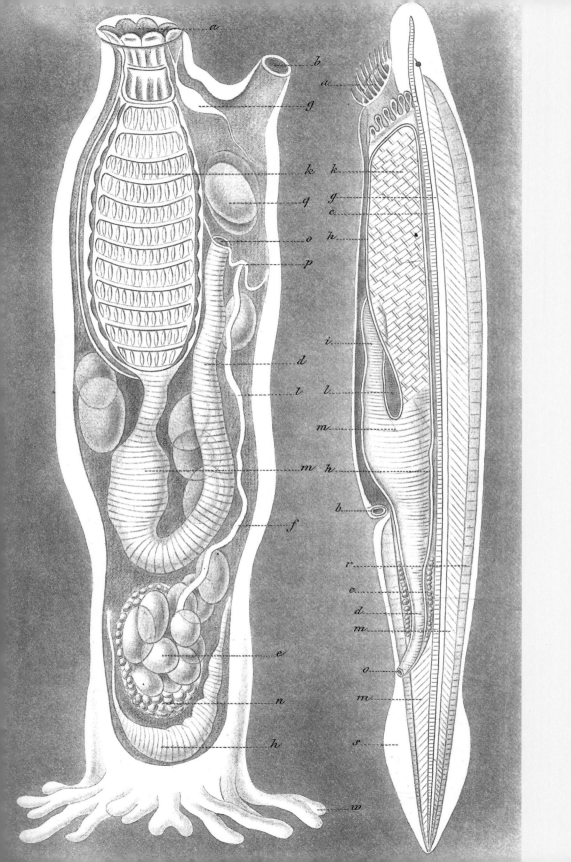

a

b

g

k   k

q   g
           c

o   h

P

d

l   l

m

m   h

f

b

r

e

e   d

           m

n

           o

h   m

           s

w

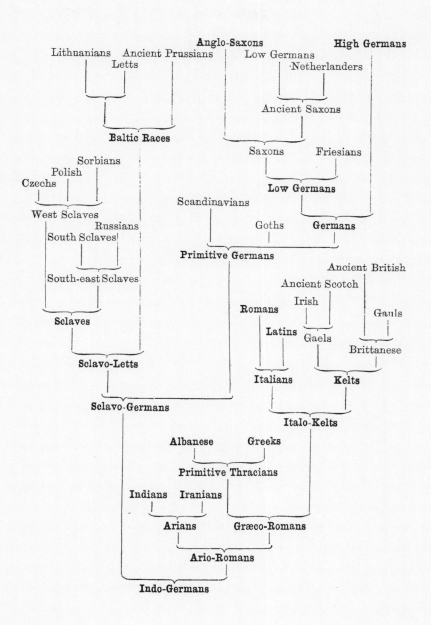

Lithuanians   Ancient Prussians
Letts

Anglo-Saxons   **High Germans**
Low Germans
·Netherlanders

Ancient Saxons

Baltic Races

Sorbians
Polish
Czechs
West Sclaves
Russians
South Sclaves
South-east Sclaves

Saxons   Friesians

Low Germans

Scandinavians
Goths   **Germans**

Primitive Germans

Ancient British

Ancient Scotch
Irish
Gauls

Romans
Latins   Gaels
Brittanese

Italians   **Kelts**

Sclaves

Sclavo-Letts

Italo-Kelts

Sclavo-Germans

Albanese   Greeks

Primitive Thracians

Indians   Iranians

Arians   **Græco-Romans**

Ario-Romans

Indo-Germans

Opposite: **Ernst Haeckel (1834–1919),** *Natürliche Schöpfungsgeschichte*, 1868; *Ascidia* (sea squirt) and Amphioxus (lancelet).

Above: **Ernst Haeckel (1834–1919),** *Anthropogenie oder Entwickelungsgeschichte des Menschen*, 1874; Pedigree of the Indo-Germanic languages.

Systematic Survey showing the derivation of the germ-layers of the Amphioxus from the parent-cell (cytula), and of the main organs from the germ-layers.

(Tree showing the ontogenetic descent of the cells in the Amphioxus).[125]

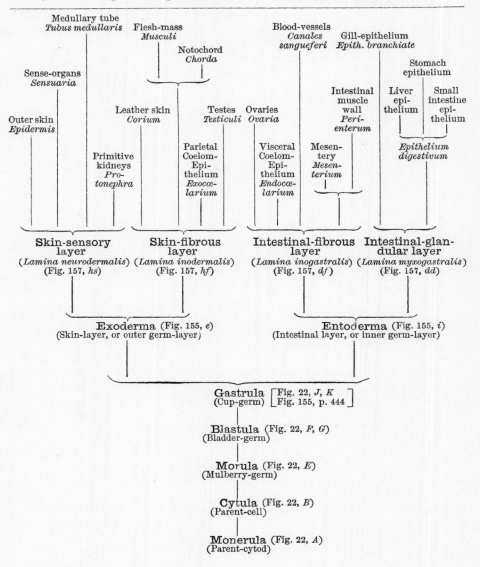

Ernst Haeckel (1834–1919), *Anthropogenie oder Entwickelungsgeschichte des Menschen*, **1874; Systematic survey showing the derivation of the germ layers of the Amphioxus.**

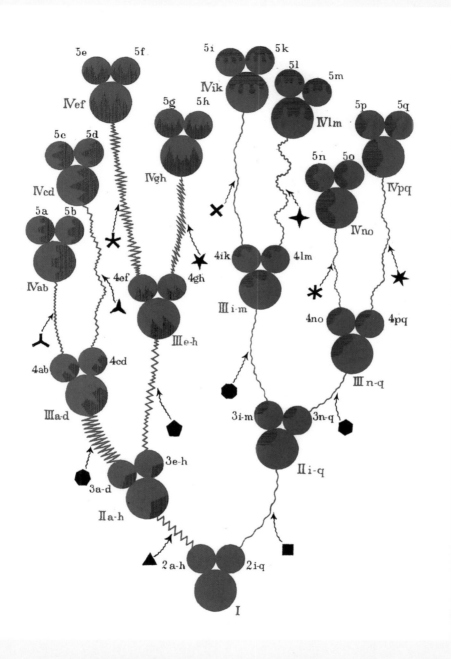

Ernst Haeckel (1834–1919), *Gemeinverständliche Vorträge und
Abhandlungen aus dem Gebiete der Entwickelungslehre*, 1902; Schema
der Perigenesis (transmission of life-force between generations).

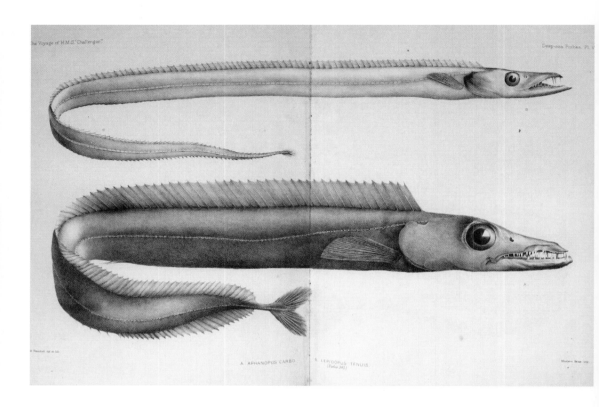

Albert Günther (1830–1914), *Report on the Deep-Sea Fishes Collected by* **H.M.S.** Challenger *during the Years 1873–1876*, **1887;** *Aphanopus carbo* **and** *Lepidopus tenuis* **at 315 fathoms.**

The *Challenger* expedition was the first global research survey of the sea, and returned a wealth of new information, especially concerning the previously unknown deeps. Indeed, the expedition sounded the deepest part of the ocean at over 35,500 feet, now known as the Challenger Deep. A century later, an ill-fated space shuttle was also named after the boat.

Albert Günther (1830–1914), *Report on the Deep-Sea Fishes Collected by* **H.M.S.** Challenger *during the Years 1873–1876*, **1887;** *Thachyrhynchus* *murrayi* **at 555 fathoms,** *T. longirostris* **and** *T. trachyrhynchus.*

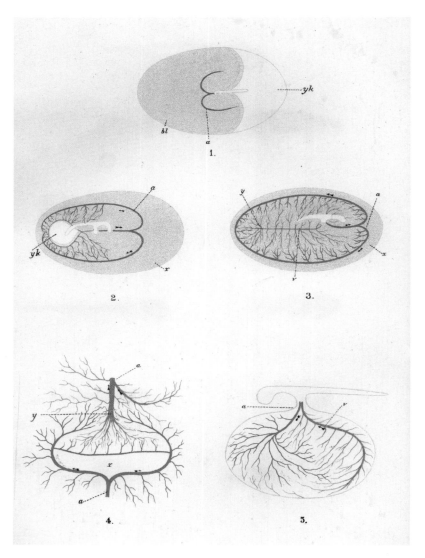

Left: Francis Maitland Balfour (1851–1882), *The Works of Francis Maitland Balfour,* **1885; Development of the yolk of** *Pristiurus* **(catshark).**

Working in Cambridge, Balfour was instrumental in discovering and comparing developmental processes in both vertebrates and invertebrates, before his untimely death in a climbing accident in the Alps.

Opposite: Adolphe Millot (1857–1921), *Le Nouveau Larousse Illustré,* **1898; Paléontologie.**

A profusion of fossil forms is presented here, but with little organization. Ancient mammoths, horse, anteaters, armadillos, fishes, turtles and molluscs jostle for prominence, often positioned according to their ability to tesselate into the smallest possible area.

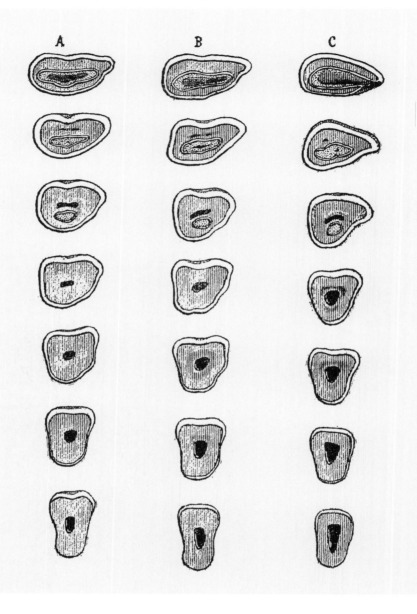

Charles Cornevin (1846–1897) and François-Xavier Lesbre (1858–1942),
*Traité De L'age Des Animaux Domestiques D'après Les Dents, 1894; Pince,
mitoyenne et coin inférieurs d'un cheval de cinq ans et demi.*

Animals' teeth may, with varying degrees of accuracy, be used to determine their
age. The age of eruption of teeth is indeed fairly consistent, but the reliability of
assessing their subsequent wear is debated. Here the pattern of wear of the lower
incisors of the horse is shown.

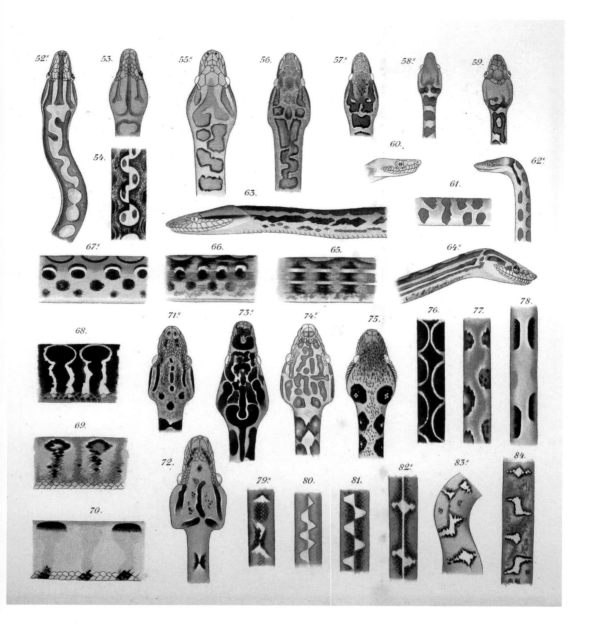

Jonathan Zenneck (1871–1959), *Die Zeichnung der Boiden,* **1898;**
Boa species from the genera *Epicrates* **and** *Corallus.*

Soon after compiling this treatise, Zenneck left biology altogether and
was to make important advances in radio and cathode-ray technology.

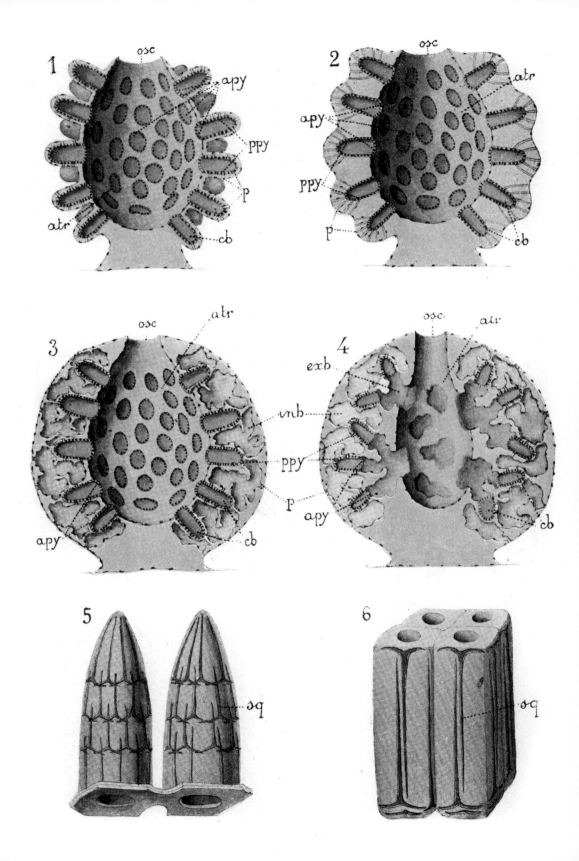

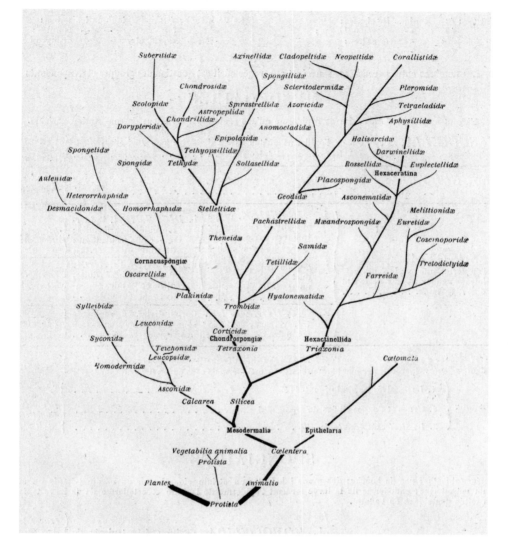

Yves Delage (1854–1920) and Edgard Hérouard (1858–1932), *Traité de Zoologie Concrète, Tome II: Mésozoaires et Spongiaires,* 1898–1899; *Arbre généalogique des spongiaires* (above); Sponges (opposite).

The sea supports a wider range of animal forms than dry land, so the classificatory challenges of its inhabitants are greater, as is their apparent alienness. In fact, some of the creatures shown opposite are closely related to the vertebrates, and thus ourselves, while others are only very distant relatives.

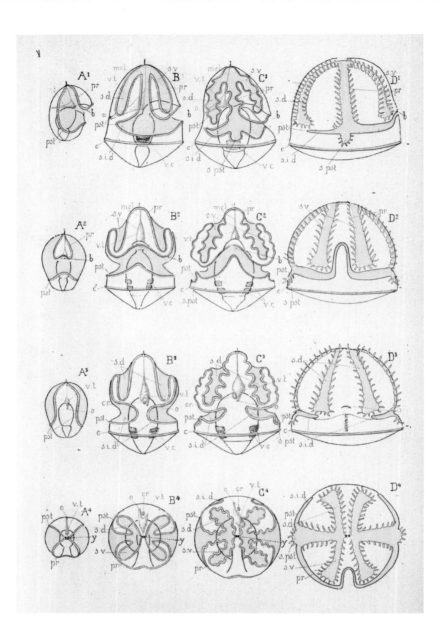

Yves Delage (1854–1920) and Edgard Hérouard, *Traité de Zoologie Concrète, Tome VIII: Les Procordés*, 1898–1899; *Balanoglossus* (above); Clavelinidae (opposite).

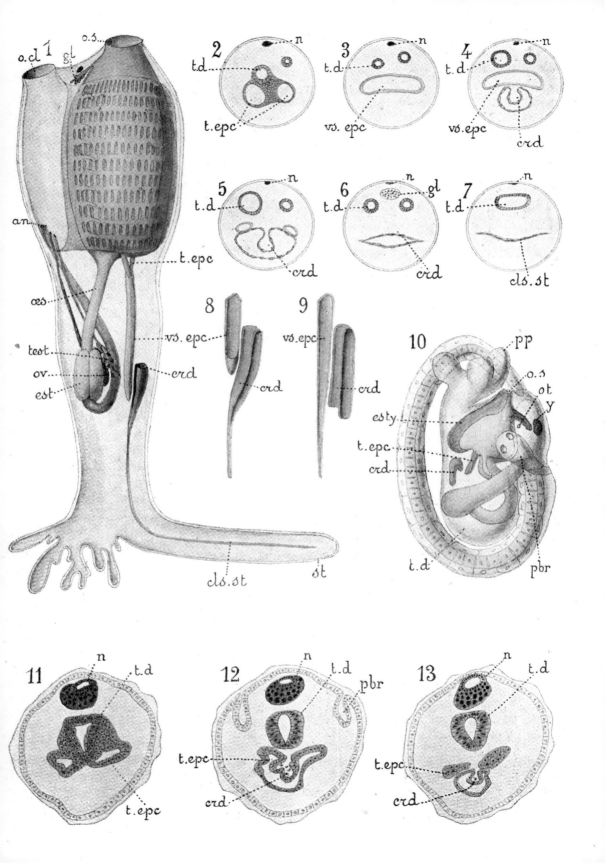

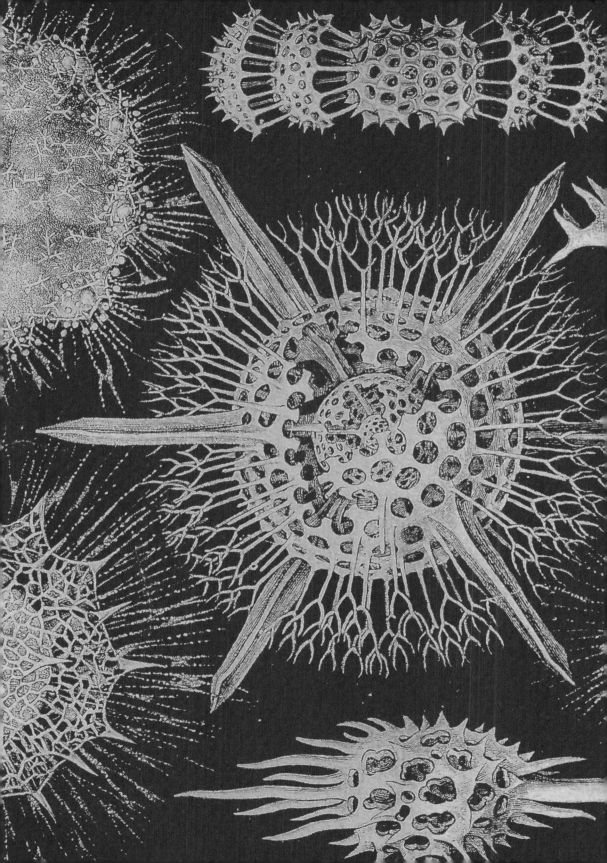

CHAPTER 4

Outward Patterns, Inner Workings

Ernst Haeckel (1834–1919), *Kristallseelen*, 1917;
Kristallinische Arbeiten der Radiolarien-Seele.

The Modern World (1900–Present)

Zoological classification has changed dramatically since 1900. There is now more information available to classify organisms. There are also more types of information and more ways to analyze it than ever before.

Because of this we must develop ever more inventive ways to graphically represent all that information. Now, not only do species change and diversify over time, but so do genes, chromosomes and genomes – twenty-four hours' sequencing of an individual creature's DNA produces far more information than can be depicted comprehensively within a single image. Also, organizing animals is no longer just about evolutionary relatedness; it also encompasses ecology, biogeography, plate tectonics, biophysics, conservation biology and many other measures by which creatures may be described, catalogued and given visual expression.

It is the field of genetics that has most changed zoology – genes, chromosomes and DNA. We are in the midst of a great upheaval in zoological classification that started with Gregor Mendel (see page 142) in the mid-nineteenth century, although his influence took many frustrating decades before it was felt. As mentioned in the previous chapter, his seminal work on the mechanisms of inheritance appeared in print just a few years after Darwin's *Origin of Species*, yet his choice to publish in the *Proceedings of the Natural History Society of Brno* meant his theories were little read until they were 'rediscovered' in the early twentieth century. Even then, it was not until the structure of DNA had been elucidated, the genetic code deciphered and DNA sequencing methods developed in the 1950s, '60s and '70s, that genes could provide the information that eager zoologists wanted to use. In the 1990s came the sequencing of single animals' entire genomes, and now we can sequence the genetic material of many individuals within a population.

It took decades to recognize the importance of Mendel's work, to discover how genetic information is inherited in the form of the DNA molecule, and to find how that information creates and controls

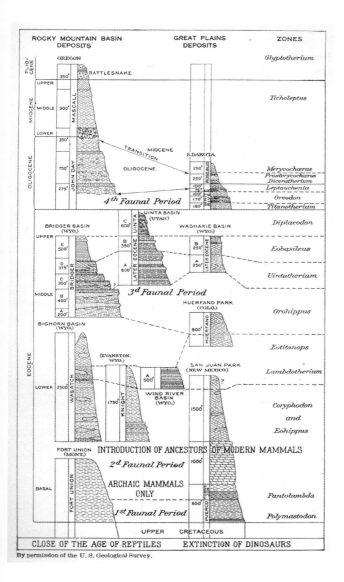

ROCKY MOUNTAIN BASIN DEPOSITS	GREAT PLAINS DEPOSITS	ZONES

(Stratigraphic chart)

PLIO-CENE — OREGON — 350' — RATTLESNAKE — *Glyptotherium*

UPPER / MIOCENE — MASCALL — 900' — *Ticholeptus*

MIDDLE

LOWER — 350'

TRANSITION — MIOCENE — S. DAKOTA

OLIGOCENE — JOHN DAY — 750' — OLIGOCENE — 250' ROSEBUD — *Merycochœrus / Promerycochœrus / Diceratherium*

275' — 250' — *Leptauchenia*

100' 25' 100' WHITE RIVER — *Oreodon*

170' 180' — *Titanotherium*

4th Faunal Period

UINTA BASIN (UTAH) — C 600' UINTA — *Diplacodon*

BRIDGER BASIN (WYO.) — WASHAKIE BASIN (WYO.) — B 350' — B 250' LATER EOCENE — *Eobasileus*

UPPER — E 500' — A 800' — A 250' — *Uintatherium*

D 375' BRIDGER — C 350'

MIDDLE — B 450' — A 200' — **3d Faunal Period** — HUERFANO PARK (COLO.) — *Orohippus*

BIGHORN BASIN (WYO.) — 600' HUERFANO — *Eotitanops*

(EVANSTON, WYO.) — SAN JUAN PARK (NEW MEXICO) — *Lambdotherium*

LOWER 2500' WASATCH — 1750' KNIGHT — A 500' — WIND RIVER BASIN (WYO.) — 1500' — *Coryphodon and Eohippus*

FORT UNION (MONT.) — **INTRODUCTION OF ANCESTORS OF MODERN MAMMALS**

2d Faunal Period — 1000'

ARCHAIC MAMMALS ONLY — 850' PUERCO TORREJON — *Pantolambda*

BASAL — FORT UNION — **1st Faunal Period** — *Polymastodon*

UPPER — CRETACEOUS

CLOSE OF THE AGE OF REPTILES — **EXTINCTION OF DINOSAURS**

By permission of the U. S. Geological Survey.

Henry Fairfield Osborn (1857–1935), *The Age of Mammals in Europe, Asia and North America*, 1921; Section through the Eocene and Oligocene formations north of lake Qûrun, Faiyum, Egypt. Arrows indicate levels richest in remains of mammals.

bodies – so in the meantime, twentieth-century biologists' organizing instincts followed two different paths. Some focused on working out those genetic details, while others continued their animal-organizing without the assistance of genes or DNA.

As early as the 1920s suspicions were growing that animal diversity was not as simple as the titans of nineteenth-century biology had believed. Surveys of embryonic sea creatures revealed that the bombastic theories of evolutionary progress put forward by Ernst Haeckel (see page 160) did not stand up to scrutiny, and analysis of rock strata showed that different parts of different animals fossilize in unexpected and capricious ways. Scientists

even interrogated Darwin and Wallace's theory of natural selection: How fast does natural selection occur? Does it proceed gradually, or in fits and starts? Is the story of evolution written by rare, grotesque mutants? Is tree-like branching the only way animals diversify? Gradually the old, staunchly rooted evolutionary trees started to look distinctly precarious.

Indeed, mid-twentieth-century evolutionary trees are very different from their gnarled, knotty predecessors. Their gently curving branches of diversification and elegant spindles of population have a neat, Art Deco flavour, and often hardly look like real trees at all. Gone is the faux-antique lettering indicating evolutionary antiquity, replaced by the sharp, clear lines of 1950's sans-serif typography. The new trees, clean and crisp, speak of careful, objective enquiry – a desire to organize the animal kingdom according to the dictates of hard-nosed twentieth-century science.

Ernst Haeckel (1834–1919), *Kristallseelen*, **1917; Frontispiece.**

Yet all this time, the actual characteristics biologists used to classify animals remained the same as those used by their Victorian forerunners. For example, the abundant dinosaur fossils excavated from the great North American fossil beds were classified according to the twists and

Max Weber (1852–1937), *Die Säugetiere*, **1927; Composition and derivation of simian dentition.**

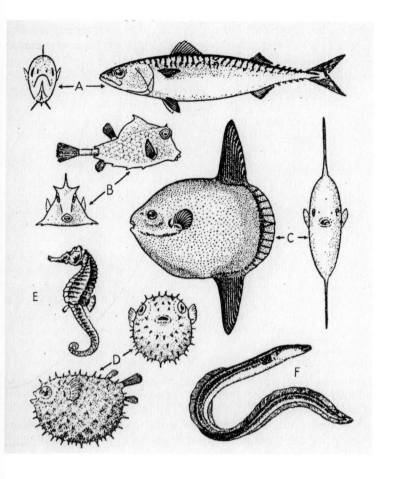

John Zachary Young (1907–1997), *The Life of Vertebrates*, 1950; Forms of fishes.

tuberosities of their bones. 'Palaeontology' means the 'study of ancient forms', and animal structure remained the key to phylogeny until surprisingly recently. For example, the long evolutionary road to the mammals was mapped largely by the examination of teeth and the tiny bones of the middle ear. Even extant species were still classified almost exclusively by their morphological characteristics. Well into the latter half of the twentieth century, anatomy – not genetics – was still king.

Soon the curvaceous clarity of the mid-century evolutionary trees was itself superseded by a new, more angular system of classification. Cladistics (from the Greek word for 'branch') is a logically simple attempt to classify species according to an algorithm-like system of simple rules, and the diagrams it yields are appropriately minimalist.

With the coming of cladistics, almost all of Linnaeus's old taxonomic gradations were discarded – the kingdom, the genus, the phylum all gone, with only the species surviving. All were replaced with a single

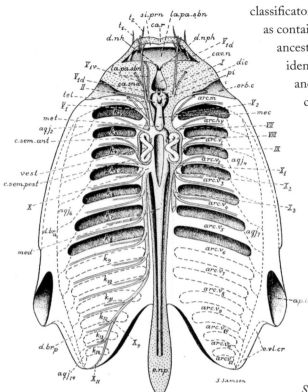

classificatory unit: the clade. A clade of organisms is defined as containing all the species descended from a single ancestor, and which have inherited some distinctive identifying shared characteristic that evolved in that ancestor. Mammals, for example, are thought to be a clade, because they are all descended from a single ancestral species that began to suckle its young. One advantage of cladistics is that it does not matter what data are used, and recently DNA sequences have proved just as useful as bony bumps to work out the genealogies of species.

It really is that simple. Even though there are clades-within-clades and continuing arguments about how clades should be identified, the same simple idea governs them all. And the visual product of all this is ascetically simple trees of bifurcating straight lines, with living species at the tips of the branches, and unknown but hypothesized grand ancestors at the root. A cladogram is as simple as a drawing of a tree can be; not unlike the single figure in *On the Origin of Species*, in fact (see page 133). It is the sort of tree Piet Mondrian might have sketched if time were short.

Pierre-Paul Grassé (1895–1985), *Traité de Zoologie, Tome XIII: Agnathes et Poissons,* **1958;** *Simopteraspis primaeva.*

Cladistics removed many assumptions from zoological classification. Evolution occurs in a branching tree pattern, but there is little sense of when or where or how the branchings took place. Also, there is absolutely no sense of progress towards perfection – on a cladogram, one animal is never superior to any other; it is simply another datum in the set. Some species might retain more ancestral characteristics, while others have inherited more new ones, but every outcome is value-neutral. There is no sense of evolutionary ascent in a cladogram – just time, increasing diversity, and the uncaring hand of natural selection. And every cladistic tree represents just a transient hypothesis that fits the available data for a while, until new data appear to dismantle and replace it.

The conceptual purity of the cladogram gives it a stark, angular beauty that appeals to the modern eye. Its simplicity gives it flexibility of form and arrangement, as in Carol Ballenger's *Tree of Life: Homage to Darwin* on page 233. Cladograms do not carry the visual gravitas of the evolutionary trees of the nineteenth century, but today's zoologist cannot help feeling that their lightness signifies genuine scientific clarity.

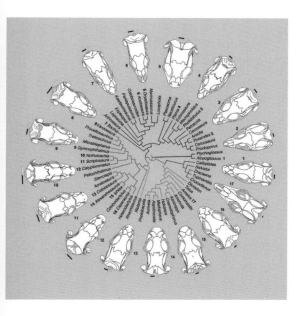

Cristian Hernandez Morales and others, Phylogenetic wheel showing variation of skull morphology among the *Gymphothalmoideagenera* (squamates), 2019, *Anatomical Record* vol. 302; Front cover.

In a further shrugging-off of biological assumptions, many contemporary evolutionary trees are not actually tree-shaped at all. Going one step beyond cladistics, 'phenetics' does not claim to establish evolutionary links between animals at all. Instead, it seeks merely to quantify similarity between species, often by a computer-enacted mathematical process. In an echo of the early-nineteenth-century zoologists who described patterns of deliberately vague 'affinity' between creatures rather than relatedness, the new forms of animal classification seem to convey only a tacit hope that similarity may one day turn out to be a sign of evolutionary relatedness.

Instead of a neatly rooted tree, recent modes of animal classification are often used to produce a network of lines, whose lengths represent the calculated dissimilarities between the species connected by those lines. The result is an irregular, rootless branching form with individual species jutting from its periphery. However, these apparently chaotic explosions of species interrelations seem to make zoologists want to think big, and many have tried to create radiating constructions connecting all known life on Earth. These giant new trees are usually based on genetic information, as DNA is one of the few things that all living organisms share. And in some of these trees there is a diminutive sprig in one corner with three tiny buds – *Homo*, *Saccharomyces* and *Zea*. Finally, we proud humans have been put in our place, relegated to a minuscule outpost on a sprawling web of organisms, along with our close cousins: yeast and sweetcorn.

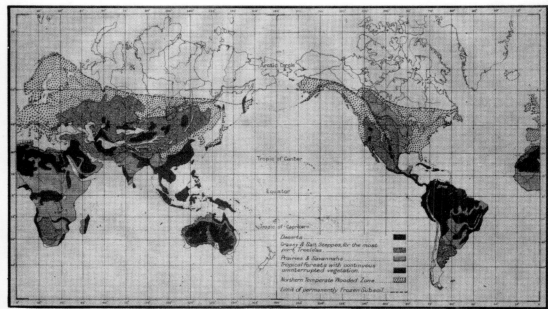

H.Gadow.

Stanford's Geog'Estab'London.

PHYSICAL FEATURES OF THE WORLD AFFECTING THE GEOGRAPHICAL DISTRIBUTION OF AMPHIBIA AND REPTILES.

Hans Friedrich Gadow (1855–1928), *Amphibia and Reptiles,* **1901; Physical features of the world affecting the geographical distribution of amphibia and reptiles (above); Skulls showing especially the composition of the bony arches of the orbito-temporal region (opposite).**

Originally from Pomerania, and working mainly in Cambridge, Gadow's zoological interests were wide, but his survey of amphibians and reptiles is one of his most widely read works. The map above shows the global distribution of the two groups – more restricted than that of mammals and birds, due to their greater dependence on environmental heat energy, and also moisture in the case of amphibians. The image opposite demonstrates the importance of windows in the side of the skull – known as 'fenestrae' – in classical taxonomy of land vertebrates.

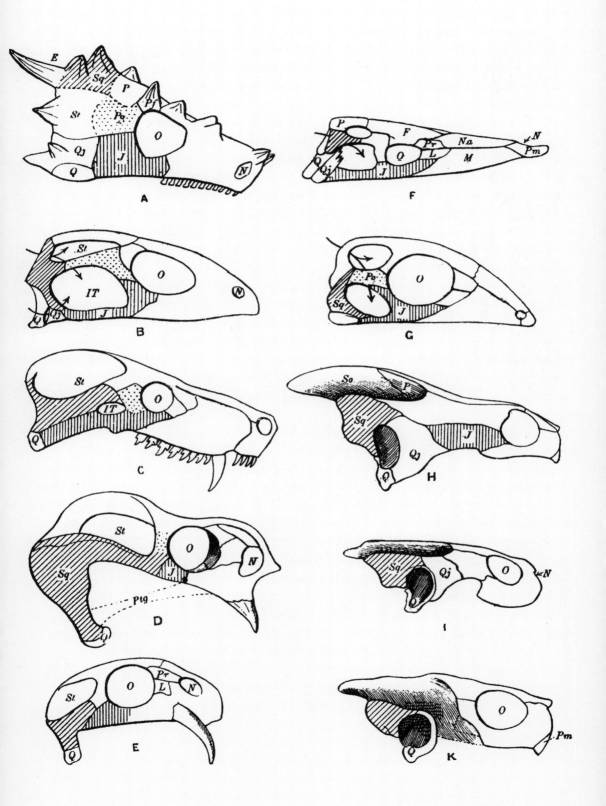

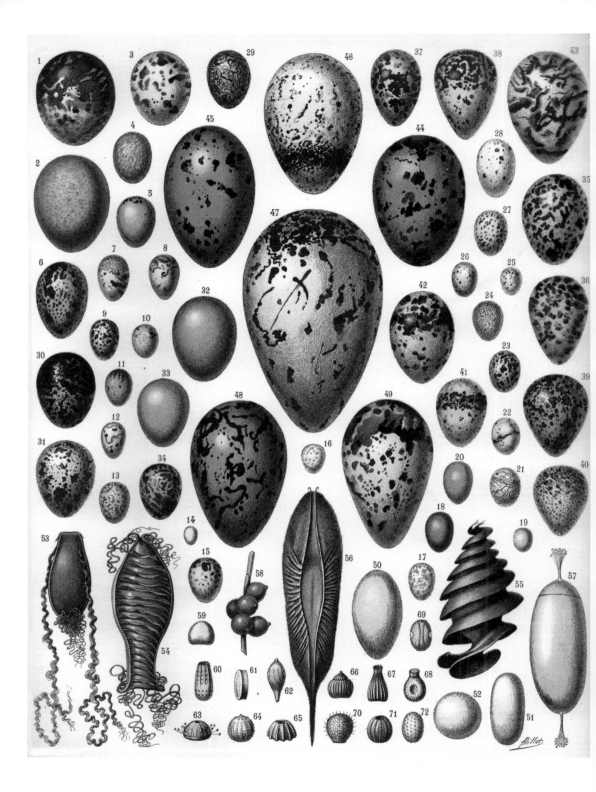

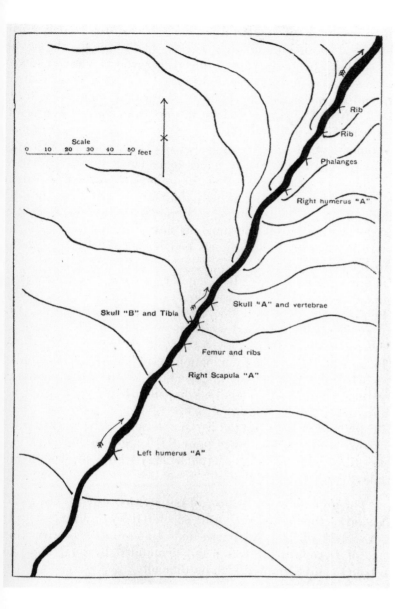

Scale

0 10 20 30 40 50 feet

Rib

Rib

Phalanges

Right humerus "A"

Skull "A" and vertebrae

Skull "B" and Tibia

Femur and ribs

Right Scapula "A"

Left humerus "A"

Left: Wynfrid Duckworth (1870–1956), *Studies from the Anthropological Laboratory, The Anatomy School, Cambridge,* **1904; The dispersive power of running water on skeletons.**

Fossilization is a selective process. Only a tiny fraction of animal remains becomes fossilized, and fossils' locations depend on environmental conditions. This image is a schematic map of a river – surface water is one of the main factors controlling the pre-fossilization transport of bones, and remarkable concentrations of a single bone type from a single species may accumulate at particular locations in a river's flow.

Opposite: Adolphe Millot (1857–1921); *Le Nouveau Larousse Illustré,* **1897–1904;** *Oeufs.*

The egg has been a biological, and indeed philosophical, conundrum since antiquity – yet often a beautiful one, as here. Unlike the microscopic oocytes of mammals, most organisms produce large eggs, which contain the nutrients required to give offspring a good start in life. Even the ova of flowering plants are now known to be the direct equivalent of the animal egg.

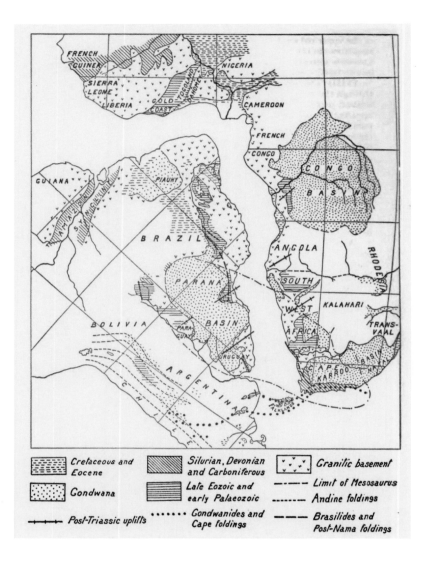

Alexander Logie du Toit (1878–1948), *A Geological Comparison of South America with South Africa*, 1927; Suggested continental restoration under the displacement hypothesis.

In the twentieth century, biology was once again united with geology, with the realization that continental drift is a major external influence affecting animal evolution and fossil preservation. Previous thinkers had noted the strange complementarity between some continental coasts that face each other across the width of an ocean, but in 1912 it was the German geophysicist Alfred Wegener who published the first coherent description of continental drift. Over subsequent decades the gradual realization that the Earth's crust consists of many plates in a state of continual movement, separation and collision was as radical to geologists as the realization of animal evolution must have been for biologists a century earlier.

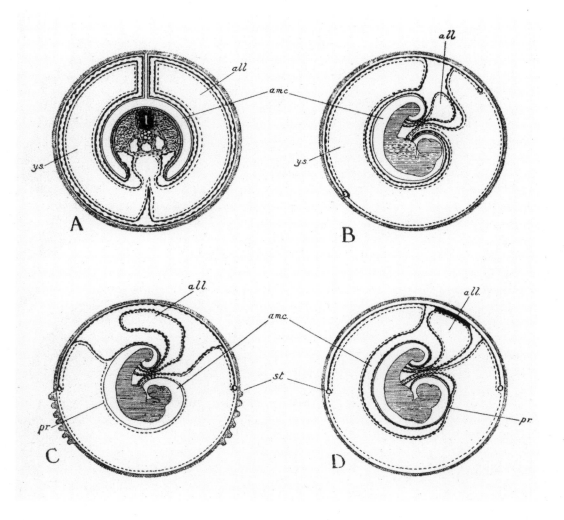

John Jenkinson (1871–1915), *Vertebrate Embryology*, 1913; Foetal membranes of monotremes (A) and three marsupials (B–D).

Vertebrates are classifiable according to their foetal membranes – the cellular linings that surround the developing embryo in the egg or *in utero*. Reptiles, birds and mammals, for example, differ from amphibians and fish in having four membranes. These membranes were also important in supporting the transition from egg-laying to live birth during the evolution of mammals – hence the analysis of monotremes and marsupials here.

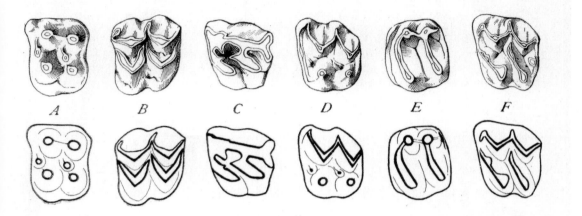

Henry Fairfield Osborn (1857–1935), *The Age of Mammals in Europe, Asia and North America*, **1921; Types of molar or grinding teeth (above); Panoramic view of the Oligocene and Miocene exposure on the south side of the White River, South Dakota (right).**

A long-serving president of the American Museum of Natural History, Osborn participated in several stratigraphical and fossil surveys of the American West – a region that, as a result, has one of the best characterized faunal timelines in the world.

Opposite: Ernst Haeckel (1834–1919), *Kristallseelen*, **1917;** *Kristallinische Arbeiten der Radiolarien-Seele.*

Haeckel continued to publish actively well into the twentieth century, although his writing became more obscure and esoteric – he even proposed his own new religion. Whatever one thinks of the written content of 'Crystal Souls: Studies of Inorganic Life', its images of the crystalline skeletons of single-celled sea creatures are as beautiful as anything in Haeckel's earlier works.

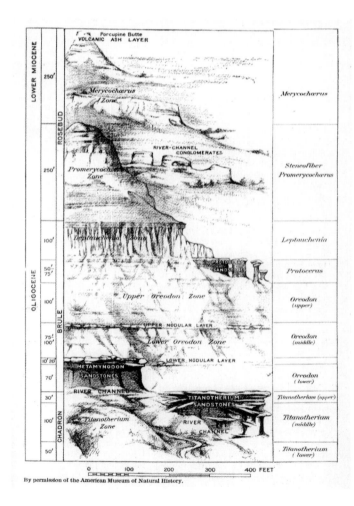

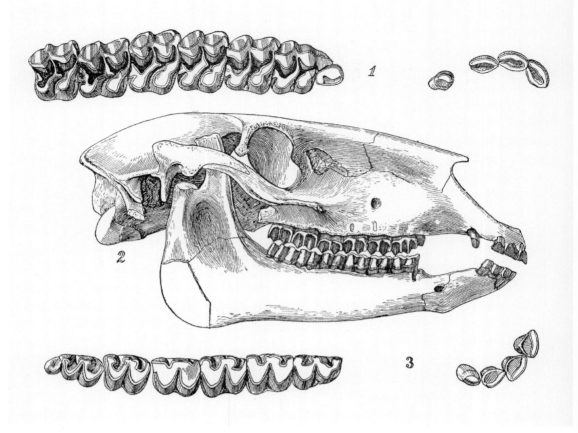

Henry Fairfield Osborn (1857–1935), *Equidae of the Oligocene, Miocene and Pliocene of North America,* **1918; Skull and dentition of type of** *Parahippus nebrascensis* **(above); The forefoot of** *Parahippus* **(opposite above); General correlation of the fossil-bearing horizons of the western plains district and of Oregon (opposite below).**

Horses have provided perhaps the most complete fossil record of any animal lineage, and palaeontologists have been able to trace not only their transition from small, dentally unspecialized, multi-toed forms to the modern *Equus*, but also the appearance, migration and demise of now-extinct branches of the equine evolutionary tree.

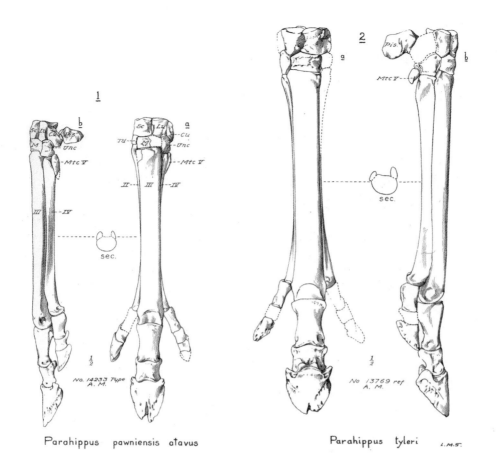

Parahippus pawniensis atavus

Parahippus tyleri   L.M.S.

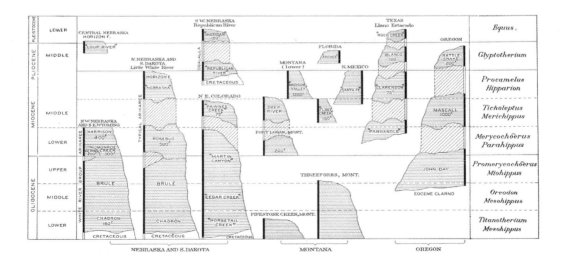

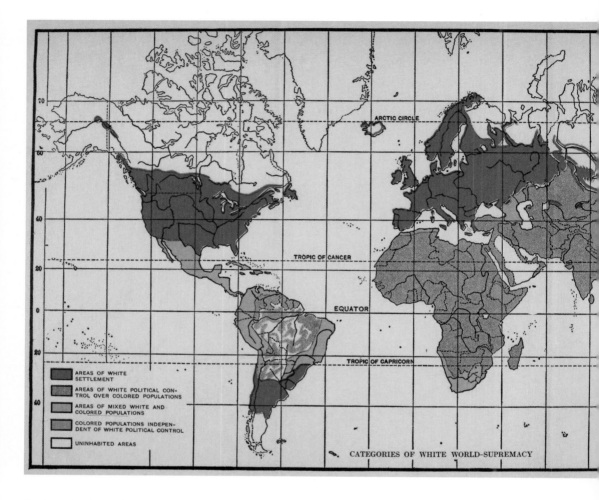

AREAS OF WHITE
SETTLEMENT

AREAS OF WHITE POLITICAL CON-
TROL OVER COLORED POPULATIONS

AREAS OF MIXED WHITE AND
COLORED POPULATIONS

COLORED POPULATIONS INDEPEN-
DENT OF WHITE POLITICAL CONTROL

UNINHABITED AREAS

CATEGORIES OF WHITE WORLD–SUPREMACY

Lothrop Stoddard (1883–1950), *The Rising Tide of Color Against
White World-Supremacy,* **1920; Categories of white world-supremacy.**

Although humans are subject to natural selective forces, the science of
evolutionary biology has often been applied incautiously to our own species.
Stoddard was an American academic whose influential ideas fed into prevailing
attitudes to racial purity, and eventually Nazi concepts of the *Untermenschen.*
This map, for example, looks superficially similar to Alfred Russel Wallace's
biogeographical maps (see pages 150–151 and 153), but the key includes categories
such as 'areas of white political control over colored populations'. The regions of
'colored populations independent of white political control', mainly China and
Japan, are coloured yellow.

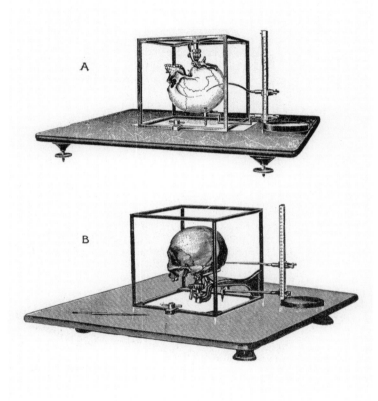

Harris Hawthorne Wilder (1864–1928), *A Laboratory Manual of Anthropometry*, **1921; Cubic craniophore.**

Anthropometry, 'measuring man', may seem an innocuous discipline, but it was often conducted with dark intentions. As would be expected for a species with such a large geographical range, humans are divided into a number of local populations, each with its own biological adaptations to its local environment. Some of these adaptations are anatomical, such as variations in skull morphology, but for many decades these differences were viewed as evidence of the superiority and inferiority of populations derived from certain parts of the world.

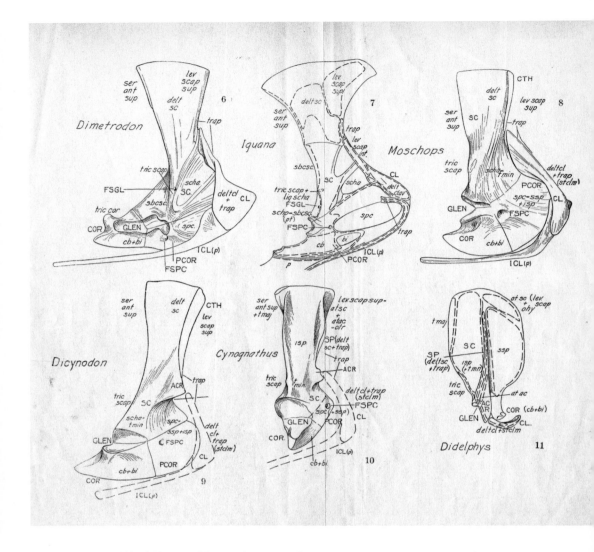

Alfred Sherwood Romer (1894–1973), *The Locomotor Apparatus of Certain Primitive and Mammal-Like Reptiles,* **1922; Forelimb girdles.**

Considerable rearrangement of the locomotor system took place in the evolutionary lineage leading to mammals. In this series, the forelimb girdle (shoulder blade, collarbone and associated structures) is compared in *Iguana*, the modern opossum *Didelphys*, and four extinct species thought to be close to the lineage leading to modern mammals. The girdle bones are seen becoming more gracile, and the attachment of muscles shifts upwards.

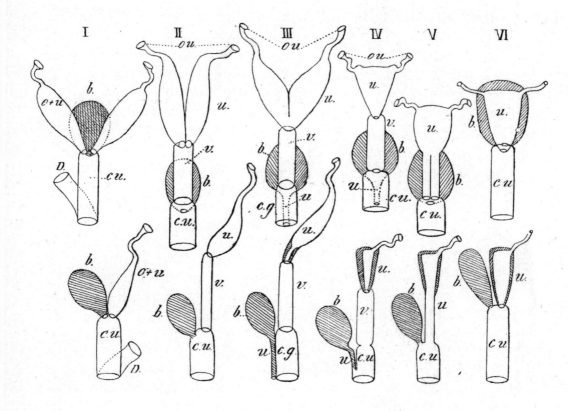

Max Weber (1852–1937), *Die Säugetiere***, 1927; Scheme of the arrangement of the uterus.**

The reproductive system exhibits dramatic variations between species, and nowhere is this more true than in marsupials. Although some species possess single uteri and cervices as humans do (IV upper row), others have uteri which are partially (III) or fully (I, II) divided into two, or have dual cervices (V). In addition, in all permutations, one fallopian tube may be absent (lower row).

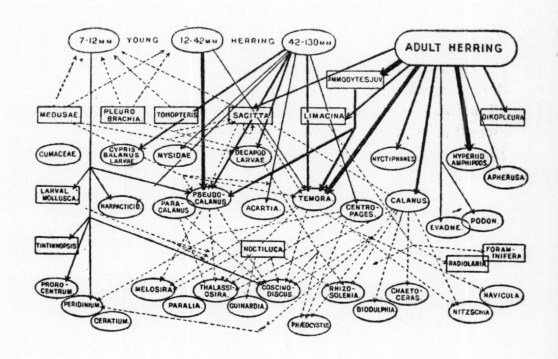

Charles Elton (1900–1991), *Animal Ecology,* **1927; The general food relations
of the herring to other members of the North Sea plankton community (above);
Fluctuations in the numbers of Canadian mammals (opposite above); Part of an
animal community in Canada (opposite below).**

'Ecology', a word coined by Ernst Haeckel (see page 160), is the study of animals'
interactions with each other and their environment, and was a new form of zoological
study largely developed in the twentieth century. The English ecologist Charles Elton
focused particularly on the flow of nutrients through environments, and developed the
modern concepts of food-chains and food-pyramids.

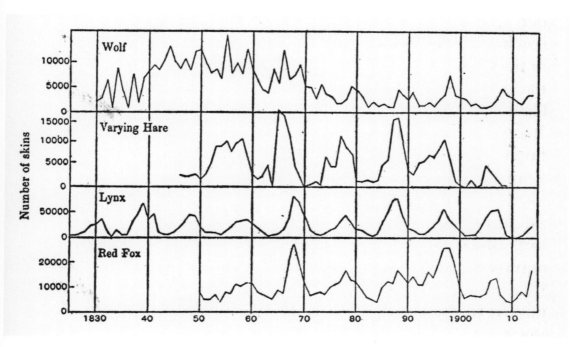

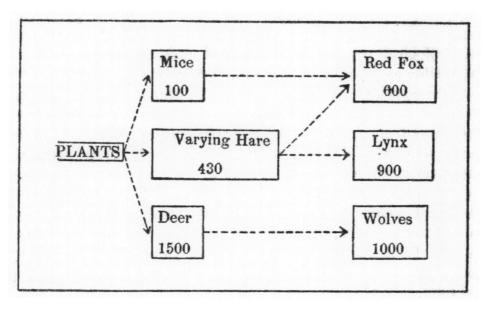

I II III IV V

VI VII VIII

Helen ZisK

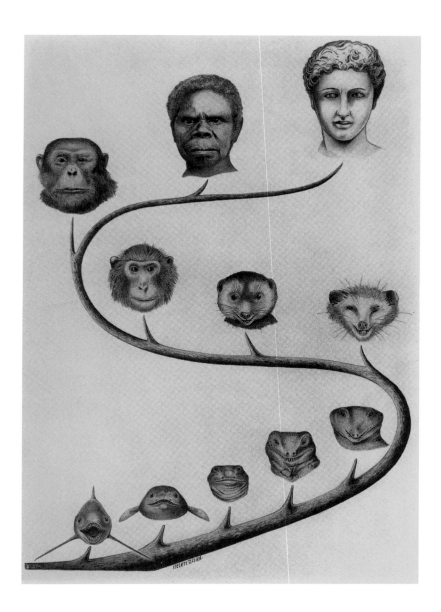

William King Gregory (1876–1970), *The Upright Posture of Man* **(illustration from a double page spread);** *Our Face from Fish to Man,* **1928.**

William Gregory was an American zoologist with particular interests in the evolution of human bipedalism, and mammalian dentition and facial structure. He was a great popularizer of science, yet his evolutionary ideas contain a sense of 'progress towards perfection' not borne out by later ideas, and which led, as here, to racist overtones.

Edwin Goodrich

Organization of the Vertebrate Body

Edwin Goodrich (1868–1946), *Studies on the Structure and Development of Vertebrates,* **1930; The urogenital system in the Craniata.**

The Oxford academic Edwin Goodrich worked at a nexus in biology, when anatomy and embryology were well-established sciences, but had not yet been changed forever by advances in genetics and molecular biology.

Goodrich travelled the world in his research, focusing especially on the organization of the body plan of diverse marine creatures. He was particularly interested in the various tubes, secretory, excretory and reproductive, that connect the inside of animals to the outside world – and the sometimes complex equivalences between these structures in different groups.

Goodrich painstakingly developed probably the most coherent pre-molecular scheme of how the vertebrate body plan evolved and how it develops in the embryo. In particular, he was able to account for gaps in the fossil record, and cope with suspicions that the evolution of the vertebrate body has not followed a satisfyingly neat path.

The diagrams, often self-drawn, selected for this book illustrate how his evolutionary/embryological schemes have influenced vertebrate morphology ever since. In one image, hidden among the complex bony mosaic of a fish's skull, lie the relatively few bones that have persisted into today's mammalian skulls. Other pictures tell the tangled evolutionary story of the vertebrate cardiovascular system, as it transitions from a one-sided system, with a single atrium and ventricle, into a two-sided system with distinct chambers pumping blood separately around the lungs and the body. Finally, the remodelling of the pelvis that took place during the evolution of dinosaurs and birds is detailed.

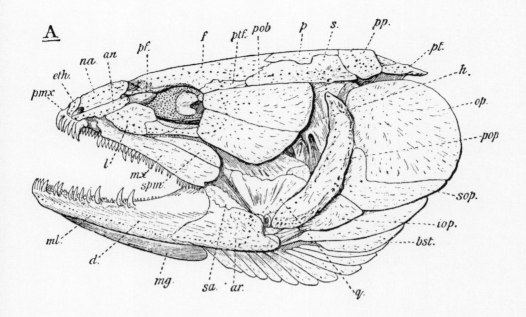

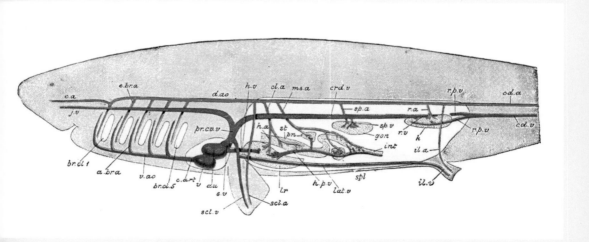

Edwin Goodrich (1868–1946), *Studies on the Structure and Development of Vertebrates*, 1930; Left-side view of the skull of *Amia calva* [bowfin] (top); Vascular system of a fish (above).

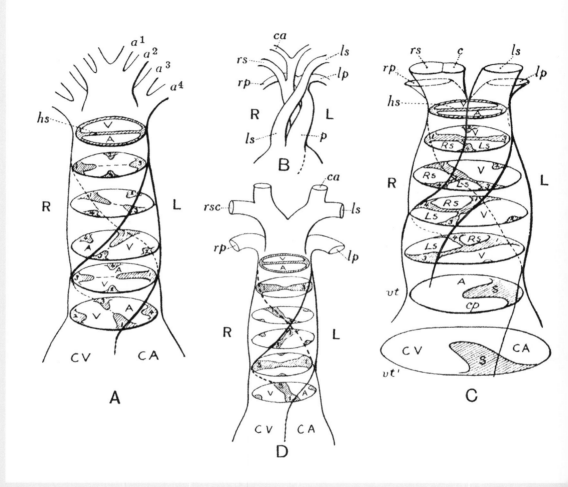

Edwin Goodrich (1868–1946), *Studies on the Structure and Development of Vertebrates*, 1930; Spiral subdivision of the bulbus cordis in Dipnoi (lungfish), Amphibia, Reptilia, Aves (birds) and Mammalia.

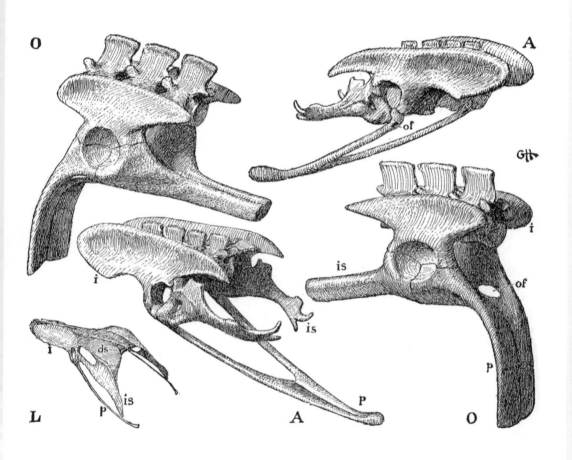

Edwin Goodrich (1868–1946), *Studies on the Structure and Development of Vertebrates*, 1930; Pelvic girdle and sacrum of A: *Archaeopteryx*, O: *Ornithosuchus* (a bipedal dinosaur), L: young gull.

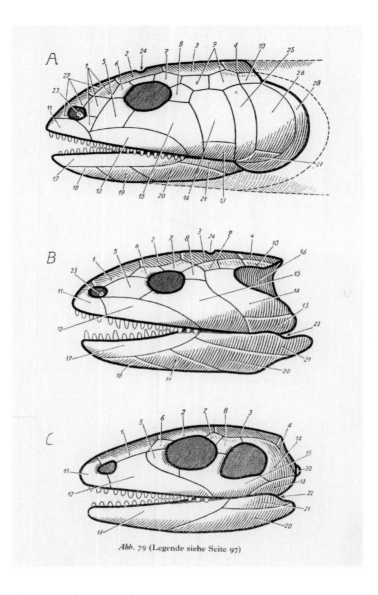

Abb. 79 (Legende siehe Seite 97)

Portmann (1897–1992), *Einführung in die Vergleichende Morphologie der Wirbeltiere*, 1948; Side view of the skulls of a crossopterygian, a stegocephalan and a primitive reptile.

Although vertebrates share an underlying body plan, this often becomes distorted over the course of evolution, and elements may be lost (but rarely added). This schematic diagram shows trends during transitions between lobe-finned fish, ancient amphibians and primitive reptiles – for example, the progressive loss of bones, lengthening of the snout, and the incorporation of holes in the side of the skull whose edges are used for the attachment of jaw muscles. The small hole for light-sensitive organs on the top of the brain (marked '24') is often lost in fully terrestrial vertebrates.

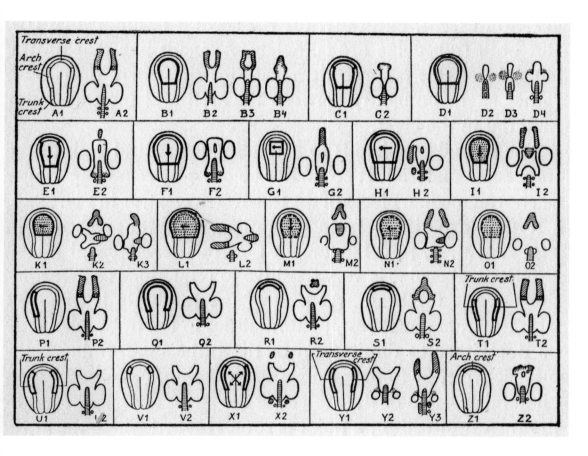

Sven Hörstadius (1898–1996), *The Neural Crest,* **1950; Development of neurocranium after extirpation, rotation or transplantation of neural plate and neural crest of the head.**

The neural crest is an embryonic cell population unique to the vertebrates, and whose evolution is thought to have conferred many of the distinctive features that have made the group so successful. It is involved in the formation of much of the skull, jaws, teeth, gills, heart valves, adrenal glands and pigment cells, and the electrical insulation of many nerve cells. This schematic table details early experiments into the role of neural crest, in which the tissue is removed, rotated or transplanted in developing embryos.

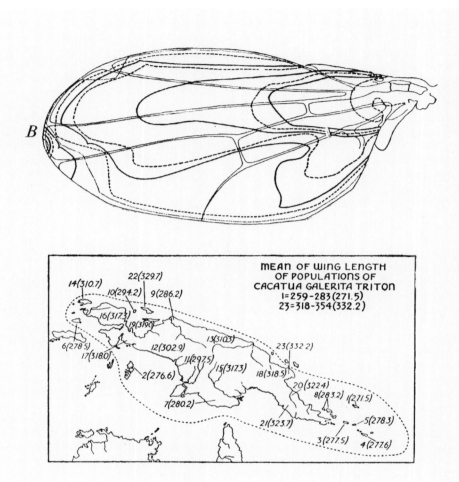

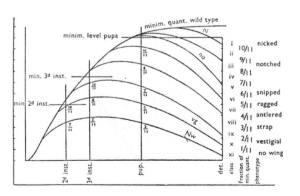

George Stuart Carter (1893–1969), *Animal Evolution*, 1951; Three stages of the development of the wing of *Drosophila melanogaster* with the 'scalloped' abnormality (top); Variation in a cockatoo in New Guinea (above); Action of the allelomorphs of 'vestigial' on the wing of *Drosophila melanogaster* (left).

Written mid-century, Carter's *Animal Evolution* incorporates the many diverse arms of evolutionary biology then established – developmental genetics, evolutionary ecology, population genetics, adaptive processes and phylogenetic trees.

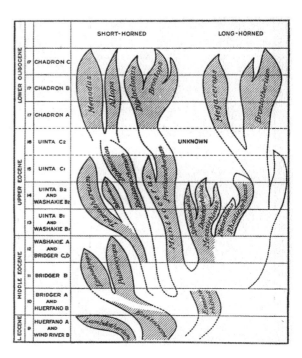

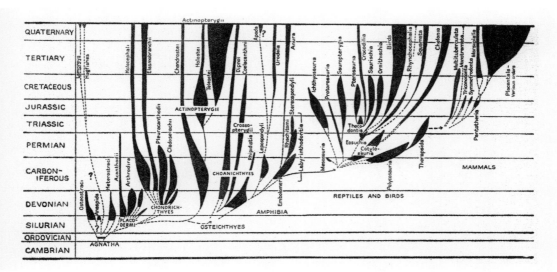

George Stuart Carter (1893–1969), *Animal Evolution*, 1951; Adaptive radiation in rays (top left); Simplified phylogeny of the titanotheres (top right); Phylogenetic tree of the vertebrates [after Romer] (above).

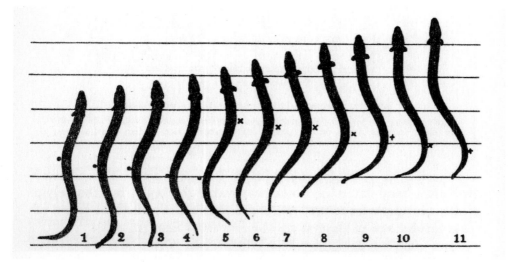

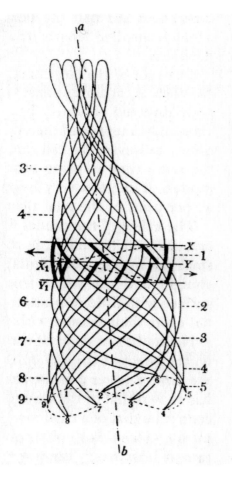

John Zachary Young (1907–1997), *The Life of Vertebrates*, 1950; Successive positions of a swimming eel (above); Drawings of successive photographs of a young eel superimposed on each other (left).

The field of biomechanics has allowed animals to be classified according to the physical processes by which they swim, fly or walk.

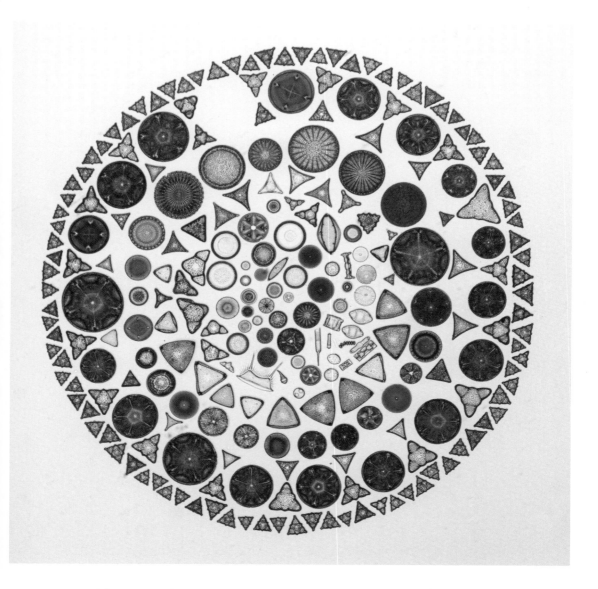

Albert Brigger (1892–1981), *Microscopic diatoms arranged on a single slide* (1952).

It is perhaps surprising that a twentieth-century scientist who spent his career studying the biology of microscopic plankton should choose to organize his subjects in such an unscientific way. Harking directly back to the microscopic montages of the Victorian era (see page 154 and 155), this artifice makes no claim to arrange these tiny creatures according to biological similarity or evolutionary relatedness. Instead, these diatoms are used simply as geometric elements in a visual whole – biological stained glass for the steady-handed.

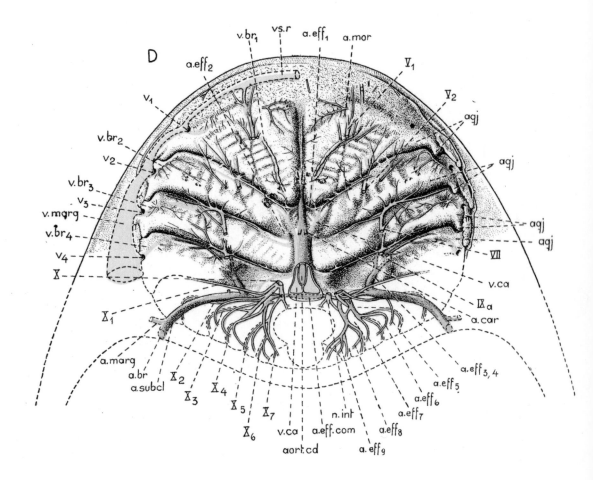

Pierre-Paul Grassé (1895–1985), *Traité de Zoologie, Tome XIII: Agnathes et Poissons,* 1958; *Zenaspis signata.*

Grassé's titanic multi-volume *Treatise on Zoology* was one of the last attempts at a single-author comprehensive survey of animal life – evolution, classification, diversity, structure, development and function. This image is a schematic of the organization of the bony carapace of one of the earliest fossil fishes. Even at this ancient date, the basic plan of the nervous system had been established, with tiny tunnels through the bone conveying nerves directly equivalent to nerves in our own heads.

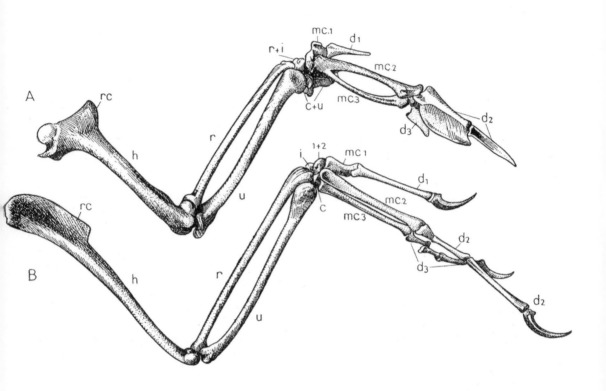

A

B

Pierre-Paul Grassé (1895–1985), *Traité de Zoologie, Tome XV: Oiseaux,*
1950; *Squelette de l'aile droite d'un pigeon* (above) *et d'Archaeornis* (below).

Birds' wings contain no 'new bones' beyond those found in their dinosaur
ancestors. However, they retain only the first three digits ('thumb', 'index' and
'middle' fingers). The lower image shows that some fossil forms still retained claws
on their wings, and that the bony fusion seen in the 'hand' of today's birds (above)
was not yet complete.

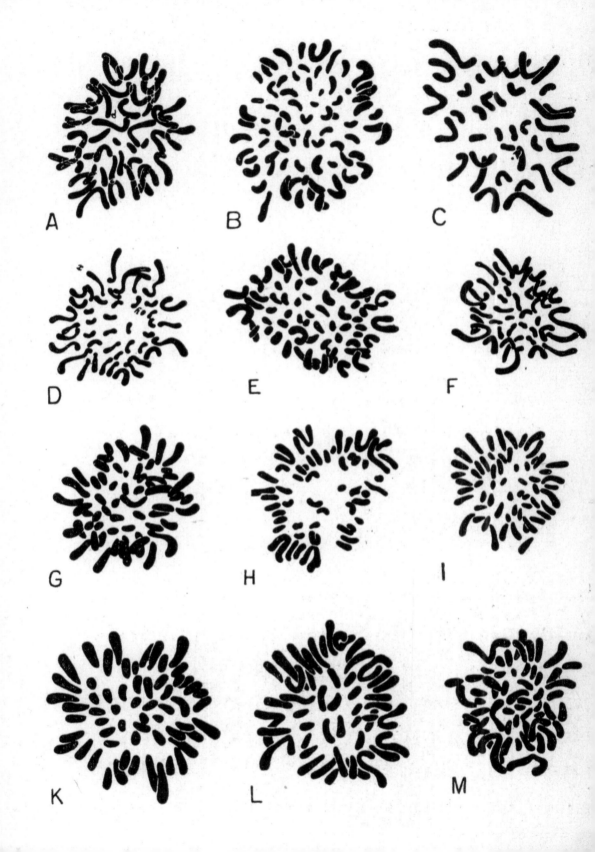

Right: Pierre-Paul Grassé (1895–1985), *Traité de Zoologie, Tome XII: Embryologie,* **1954; Brain of a bird (top); Brain of a mammal (bottom).**

The brain is the most complex organ in the vertebrate body, yet it displays a common pattern of organization in all vertebrates, including humans. There are variations between different vertebrate groups, but these are often due to differences in preponderance of brain regions present in some form in all of them.

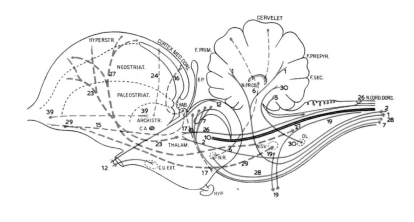

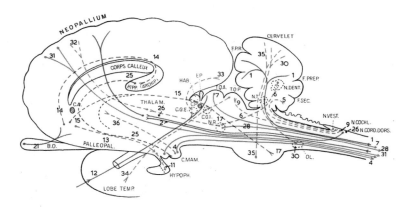

Opposite: Pierre-Paul Grassé (1895–1985), *Traité de Zoologie, Tome XII: Embryologie,* **1954; Metaphase diploid chromosomes of twelve mammals.**

It was known in the eighteenth century that at a certain phase of cell division, now called 'metaphase', a clump of darkly stained strands can be seen in microscopic preparations. These 'coloured bodies', or 'chromo-somes', were suspected to contain the cell's genetic material long before the discovery of the double-helix structure of their constituent DNA molecule in 1953.

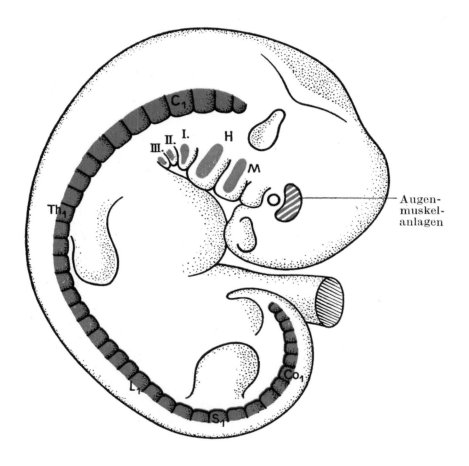

Augen-
muskel-
anlagen

Dietrich Stark (1908–2001), *Embryologie,* **1955;** *Schema der Muskelanlagen, menschlicher Embryo von 5mm Länge* **(above);** *Schema der wichtigsten Muskelgruppen beim älteren Embryo* **(opposite).**

Our evolutionary heritage is obvious during the development of a human baby. In these two images, almost all muscles of the human body are seen to be derived from two different segmental precursors. The red segments are myotomes – equivalent to the stripes of muscle in fish fillets – and they form the muscles of the trunk, limbs and tongue, as well as the unusual muscles which move the eye (*Augenmuskel-anlagen*). The blue segments are the pharyngeal arch muscles – evolved from the muscles that moved our fishy ancestors' gills – which now form most other muscles in the head and neck.

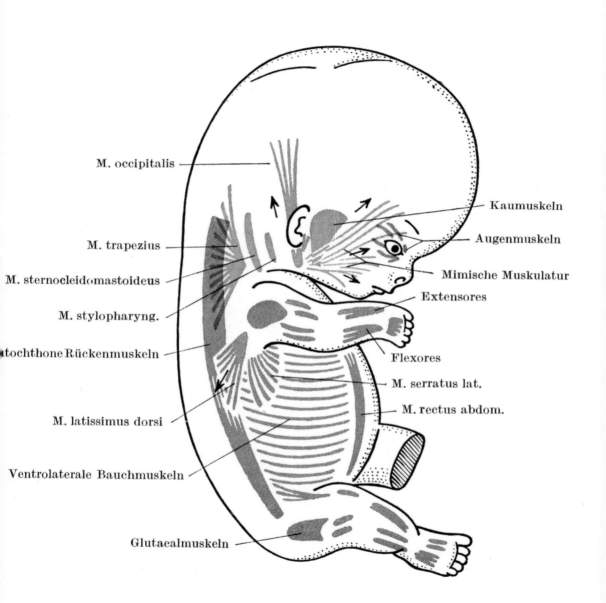

M. occipitalis

Kaumuskeln

M. trapezius

Augenmuskeln

M. sternocleidomastoideus

Mimische Muskulatur

M. stylopharyng.

Extensores

tochthone Rückenmuskeln

Flexores

M. serratus lat.

M. latissimus dorsi

M. rectus abdom.

Ventrolaterale Bauchmuskeln

Glutaealmuskeln

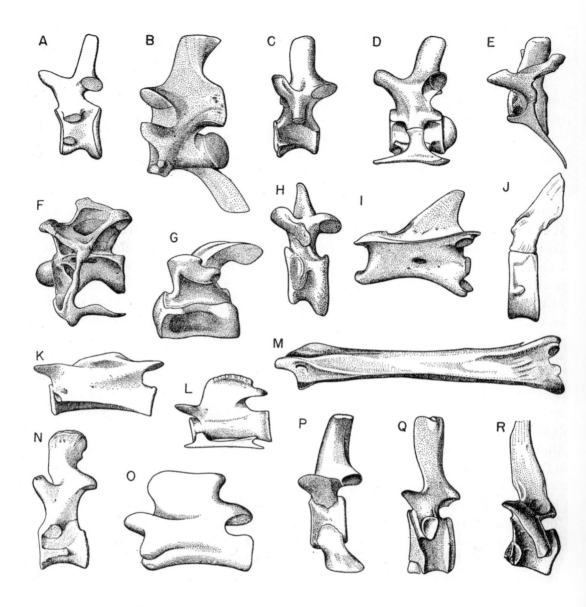

Alfred Sherwood Romer (1894–1973), *Osteology of the Reptiles,* **1956; Cervical vertebrae.**

Whether knowingly or not, many of our images of vertebrates, and especially those of extinct reptiles, are based on work by Romer. His meticulous analyses of reptilian skeletons created the armature on which most subsequent reconstructions of these captivating creatures were built.

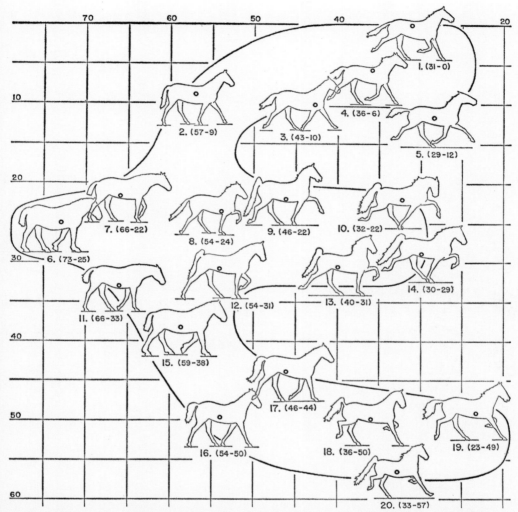

Fig. 6. In the background is shown the area of the basic graph within which fall nearly all gait formulas for symmetrical gaits of horses. Twenty specific formulas are located (small circles), and around each is drawn a silhouette of the horse moving as represented by the formula. In every sketch the left hind foot has just touched the ground.

Milton Hildebrand (b. 1918), *Symmetrical gaits of horses,* **1965,** *Science* **vol. 150, pp. 701–708.**

The natural gaits of the horse, and indeed many other quadrupeds, are the walk, the trot, the canter and the transverse gallop. However, having four legs makes possible a bewildering variety of gaits, which have proved remarkably difficult to define and classify.

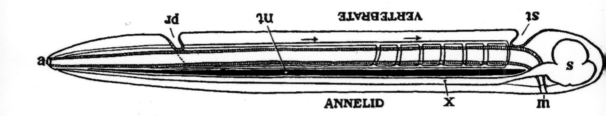

Alfred Sherwood Romer (1894–1973), *The Vertebrate Body*, **1970; The similarities between the vertebrate body plan and an inverted annelid.**

One of the most far-reaching attempts to organize, or perhaps over-organize, the animal kingdom started in the early eighteenth century. It was realized that the structure of vertebrates was quite unlike that of many animals, such as insects and segmented worms. However, the French biologist Geoffroy Saint-Hilaire thought he had solved the problem with the startling suggestion that the common vertebrate ancestor had once indeed resembled those other creatures, but had simply started swimming *upside down*. The simplistic nature of this theory meant it was largely ignored for over a century, until some (but not all) modern developmental studies provided support for it. The theory remains contentious, if tenacious, so it is not yet clear whether vertebrates truly are worms that 'turned over'.

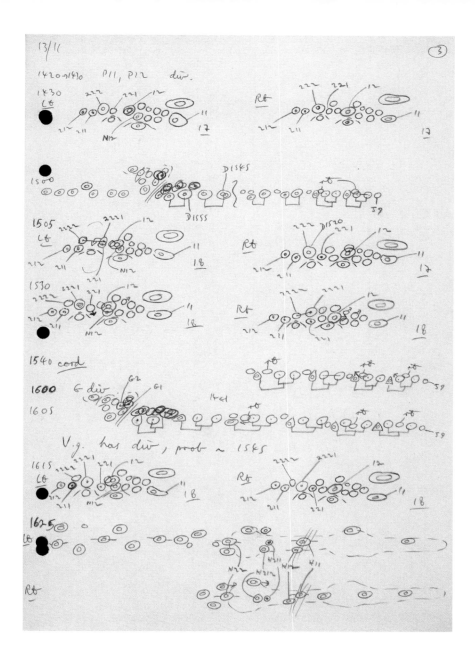

Wellcome Collection, London, *Notebook page*; **Coloured cell lineage drawings for the** *Caenorhabditis elegans* **worm tail.**

One of the most astounding feats of zoological investigation was the determination of the precise pattern of division and specialization of every cell in the larva of a multicellular organism, the roundworm *Caenorhabditis elegans*. A total of 671 cells are unfailingly produced during this process, and these sketches show the research team's attempts to elucidate the "family tree" of the cells in just one part of the body. As a result of these studies, this humble creature has become a mainstay of modern developmental biology.

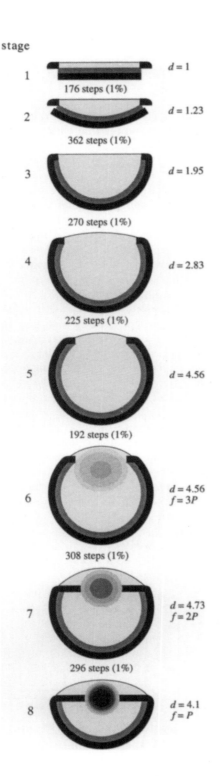

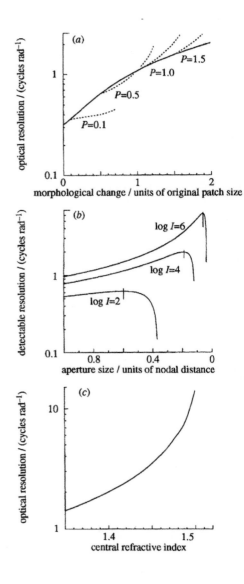

Dan Nilsson and Suzanne Pelger, A pessimistic estimate of the time required for an eye to evolve, 1994, *Proceedings of the Royal Society B* vol. 256, pp. 53–58.

Even Darwin worried about how something as apparently perfect as the eye could evolve by natural selection rather than by divine design. However, more recent simulations of the process of eye evolution suggest that it can take place remarkably rapidly, which may explain why so many disparate animals possess eyes.

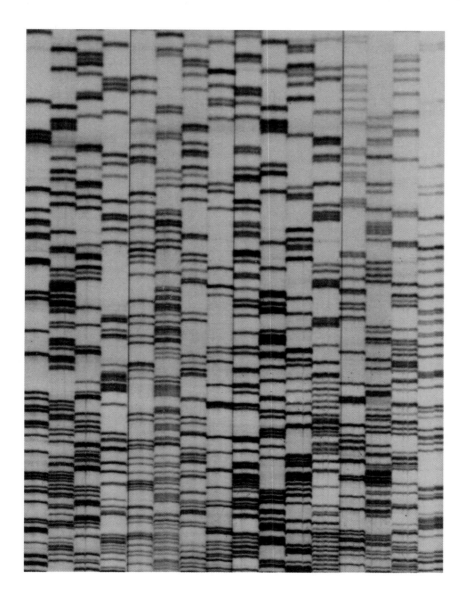

David Bainbridge, *Sequences of transplantation antigen complementary DNAs derived from invasive bovine trophoblast cells,* **1999.**

The ability to determine the sequence of DNA has transformed experimental biology – not only can scientists observe the structure, function and behaviour of animals, but they can now also decrypt their underlying genetic code. This image is from the author's own research into a particular set of genes postulated to have a role in immunity during pregnancy in ruminant animals, using the now-superseded technique of radioactive DNA sequencing. The DNA code consists of sequences of four possible bases named A, C, G and T – each set of four vertical columns is the sequence of a single nucleic acid molecule purified from a cell sample.

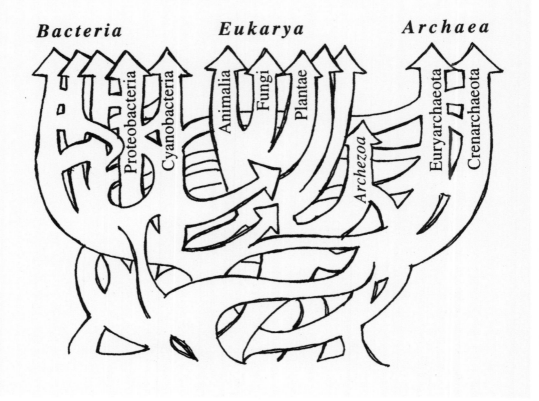

W. Ford Doolittle, *Phylogenetic classification and the universal tree,* **1999,**
Science **vol. 284, pp. 2124–2128.**

By the turn of the twenty-first century it was clear that organisms do not evolve
in a simple, branching, tree-like pattern. First, evidence began to accumulate
that much of the DNA in many bacteria seems to have been acquired remarkably
recently, apparently purloined from other bacteria by a process of 'lateral transfer'.
Second, viruses are thought to provide a route for passing extensive regions of
DNA from one complex animal (such as a snake) to another (such as a cow).
Finally, it is now accepted that the complicated 'eukaryote' nucleated cells that
make up animals, plants and fungi are actually conglomerations of at least two,
and probably more, ancestral bacteria. As a result of all this sharing and merging,
and as the images on these two pages show, the 'tree of life' has changed radically
– not only can one branch now split into two, but two branches can, confusingly,
merge into one.

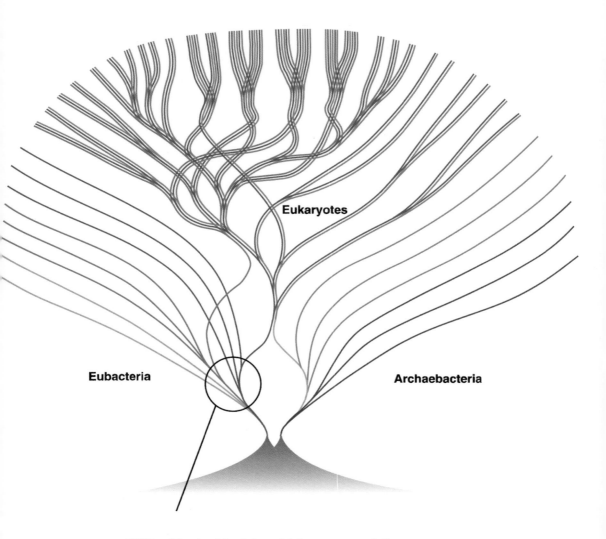

Eukaryotes

Eubacteria

Archaebacteria

William Martin, *Mosaic bacterial chromosomes: a challenge en route to a tree of genomes*, 1999, *Bioessays* vol. 21, pp. 99–104.

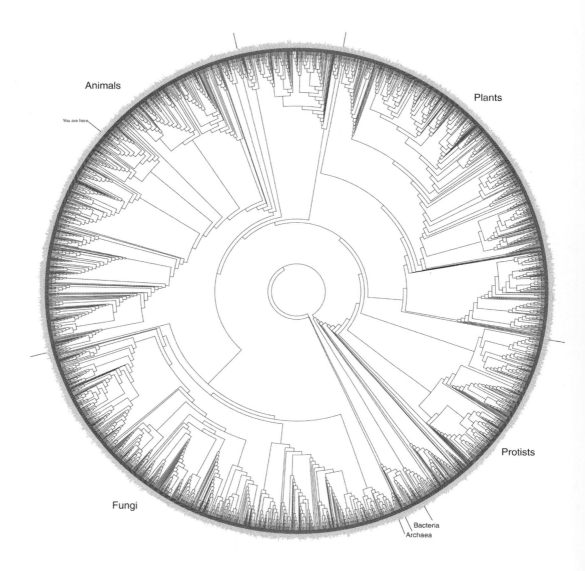

Animals

You are here

Plants

Protists

Fungi

Bacteria
Archaea

David Hillis and Elizabeth Pennisi, *Modernising the Tree of Life,* **2003,**
Science **vol. 300, pp. 1692–1697; Hillis Plot.**

The so-called 'Hillis Plot' is a striking attempt to produce an evolutionary tree for
all living things. Because the terminal twigs on any tree greatly outnumber the major
branches, many phylogeneticists now reformat their trees as circles, with a central
'trunk', and terminal twigs at the circumference, as here. The fuzzy 'rind' of the Hillis
Plot circle is actually the names of the constituent species written in a minuscule
font. The Hillis Plot is notable for its skew towards eukaryotic organisms (those
with nucleated cells such as fungi, plants and animals), while the far more numerous
bacteria and archaea are relegated to a tiny slice at around five o'clock. Some iterations
of the plot feature the tiny announcement 'you are here' at half past ten.

Carol Ballenger, *Tree of Life: Homage to Darwin,* **2007.**

A gentle play on words, the centre of this image features a transected tree trunk, on which is superimposed the Hillis Plot.

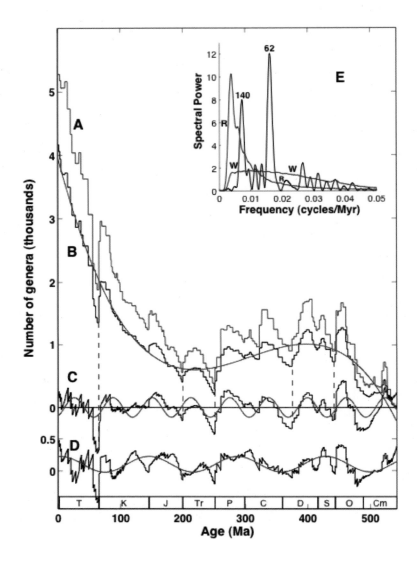

Robert Rohde and Richard Müller, *Cycles in fossil diversity*, 2005,
Nature vol. 434, pp. 208–210.

A complex graph with an unnerving interpretation. The green line 'A' represents
the number of known marine fossil groups present over geological time, from
the present on the left, to 500 millions years ago on the right. The data are then
neatened up in traces B, C and D, and some trends removed, to provide evidence
that major extinction events occur with a regular interval of 62 million years.
The causes of this postulated cycle are unknown, but may be celestial.

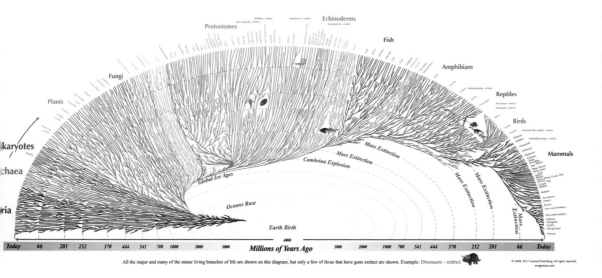

All the major and many of the minor living branches of life are shown on this diagram, but only a few of those that have gone extinct are shown. Example: Dinosaurs - extinct

L. Eisenberg, *Tree of Life* **(evogeneao.com).**

A teaching tool designed to promote the teaching of evolution, the Evogeneao tree solves the one-trunk-many-branches design problem by employing a half-ellipse format.

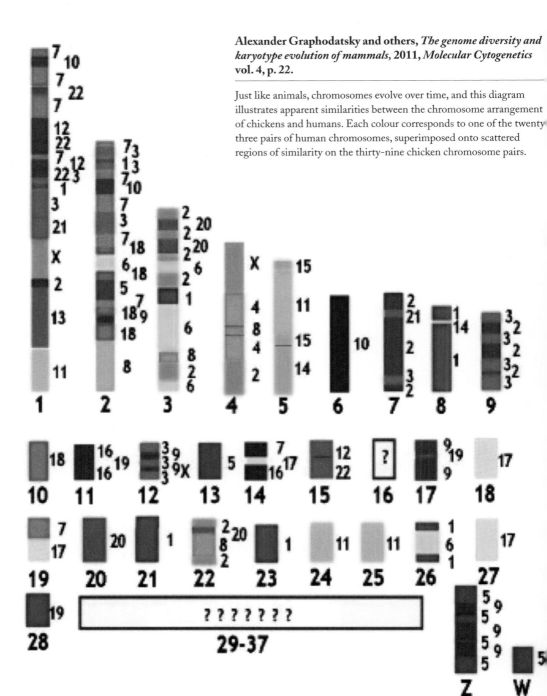

Alexander Graphodatsky and others, *The genome diversity and karyotype evolution of mammals,* **2011,** *Molecular Cytogenetics* **vol. 4, p. 22.**

Just like animals, chromosomes evolve over time, and this diagram illustrates apparent similarities between the chromosome arrangement of chickens and humans. Each colour corresponds to one of the twenty three pairs of human chromosomes, superimposed onto scattered regions of similarity on the thirty-nine chicken chromosome pairs.

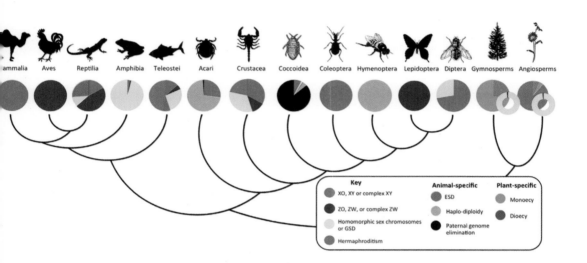

The pie charts and phylogenetic tree show labels: ammalia, Aves, Reptilia, Amphibia, Teleostei, Acari, Crustacea, Coccoidea, Coleoptera, Hymenoptera, Lepidoptera, Diptera, Gymnosperms, Angiosperms.

Key

XO, XY or complex XY

ZO, ZW, or complex ZW

Homomorphic sex chromosomes or GSD

Hermaphroditism

Animal-specific

ESD

Haplo-diploidy

Paternal genome elimination

Plant-specific

Monoecy

Dioecy

Doris Bachtrog, *Sex determination: why so many ways of doing it?*, 2014, *PLOS* **article 1001899.**

The mechanisms by which animals determine their sex during embryonic development are remarkably diverse. Mammals use a system based on the X and Y sex chromosomes (red in the figure), as indeed do beetles (*Coleoptera*). In contrast, birds (*Aves*, blue) use a chromosomal Z/W system, which appears superficially to be the opposite of the mammalian system (ZZ birds are male, ZW are female). Butterflies and moths (*Lepidoptera*) use a bird-like system. Reptiles use a variety of systems, including environmental sex determination (green) in which the sex of offspring is controlled by the temperature at which their eggs are incubated.

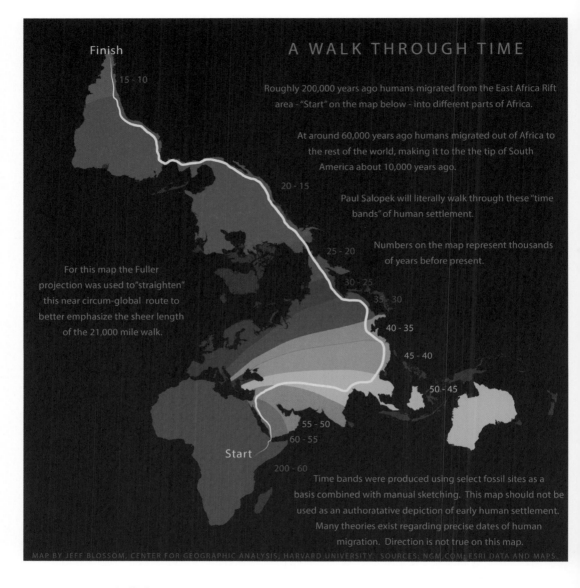

A WALK THROUGH TIME

Roughly 200,000 years ago humans migrated from the East Africa Rift area – "Start" on the map below – into different parts of Africa.

At around 60,000 years ago humans migrated out of Africa to the rest of the world, making it to the the tip of South America about 10,000 years ago.

Paul Salopek will literally walk through these "time bands" of human settlement.

Numbers on the map represent thousands of years before present.

For this map the Fuller projection was used to "straighten" this near circum-global route to better emphasize the sheer length of the 21,000 mile walk.

Finish

15 - 10

20 - 15

25 - 20

30 - 25

35 - 30

40 - 35

45 - 40

50 - 45

55 - 50

60 - 55

Start

200 - 60

Time bands were produced using select fossil sites as a basis combined with manual sketching. This map should not be used as an authoratative depiction of early human settlement. Many theories exist regarding precise dates of human migration. Direction is not true on this map.

MAP BY JEFF BLOSSOM, CENTER FOR GEOGRAPHIC ANALYSIS, HARVARD UNIVERSITY. SOURCES: NGM.COM; ESRI DATA AND MAPS.

Jeff Blossom, *A Walk Through Time*, 2015.

Humans are the most widespread of all mammalian species, yet their geographical spread has been remarkably recent and rapid. This image is an 'isochrone map', in which the range of the human species at particular times is shown by different colours. The unusual map projection allows all continents to be depicted with relatively little distortion to their intrinsic shape.

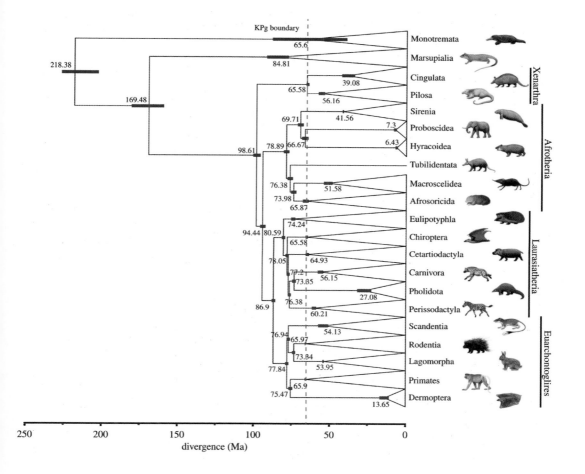

KPg boundary

Monotremata	
Marsupialia	
Cingulata	Xenarthra
Pilosa	Xenarthra
Sirenia	Afrotheria
Proboscidea	Afrotheria
Hyracoidea	Afrotheria
Tubilidentata	Afrotheria
Macroscelidea	Afrotheria
Afrosoricida	Afrotheria
Eulipotyphla	Laurasiatheria
Chiroptera	Laurasiatheria
Cetartiodactyla	Laurasiatheria
Carnivora	Laurasiatheria
Pholidota	Laurasiatheria
Perissodactyla	Laurasiatheria
Scandentia	Euarchontoglires
Rodentia	Euarchontoglires
Lagomorpha	Euarchontoglires
Primates	Euarchontoglires
Dermoptera	Euarchontoglires

divergence (Ma)

Nicole Foley and others, *Mammal madness: is the mammal tree of life not yet resolved?*, **2016,** *Philosophical Transactions of the Royal Society B* **vol. 371, article 20150140.**

The evolutionary relationships of the twenty-or-so groups of mammals have proved surprisingly difficult to determine. Many of these groups diverged from each other a long time ago, but relatively rapidly, and have since followed a bewildering variety of evolutionary paths – a combination of circumstances that probably means their genealogy will remain unresolved for some time into the future.

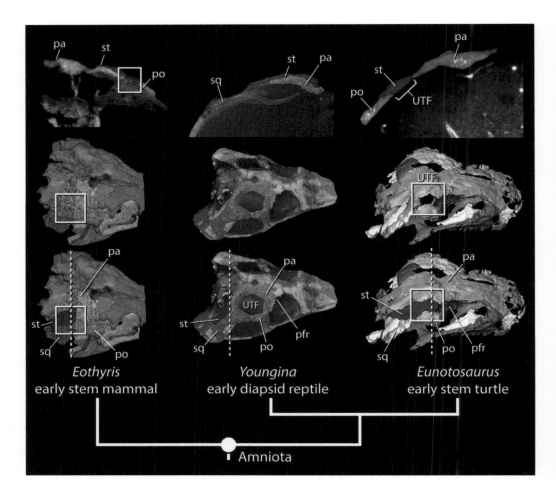

Gabriel Bever and others, *The amniote temporal roof and the diapsid origin of the turtle skull*, 2016, *Zoology* vol. 119, pp. 471–473.

For decades, the 'amniote' vertebrates (reptiles, birds and mammals) have been classified according to the windows or 'fenestrae' in the sides of their skulls – with turtles having no fenestrae, mammals having one, and birds and other reptiles having two. It was long assumed that turtles retained the ancestral arrangement, with the result that it was not possible to determine which of these three groups were most closely related. However, recent studies using micro-computerized tomography (µCT) scans have shown that the 'windowless' turtle skull is actually derived from a 'two-window' ancestor rather than the other way round. Thus, the pattern of divergence of the three dominant groups of land vertebrates seems now to have been resolved, with mammals only distantly related to all reptiles and birds.

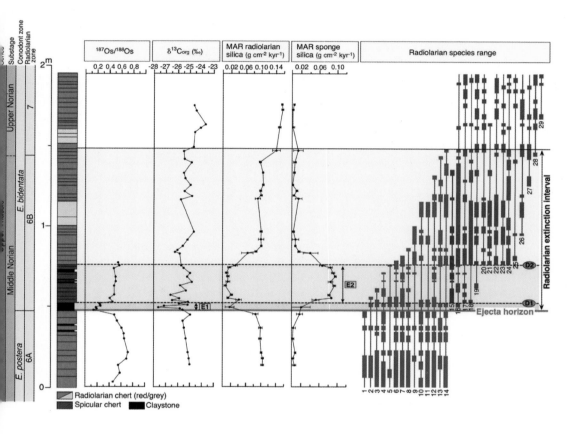

Tetsuji Onoue and others, *Bolide impact triggered the late Triassic extinction event in equatorial Panthalassa,* **2016,** *Nature Scientific Reports* **article 29609.**

Rather than appearing only as instantaneous events in the fossil record, catastrophic extinctions can now be tracked through geological time. This graph traces marine sediment chemical isotopes, species diversity, and animal-origin silica production over a relatively brief period around the impact, 201 million years ago, of a large asteroid. Moving upwards from the lower edge of the graph, the time of impact is marked by the horizontal red line, after which there is a period of 300,000 years when the various measures of ecosystem health are dramatically perturbed. There was a simultaneous extinction of many land animals.

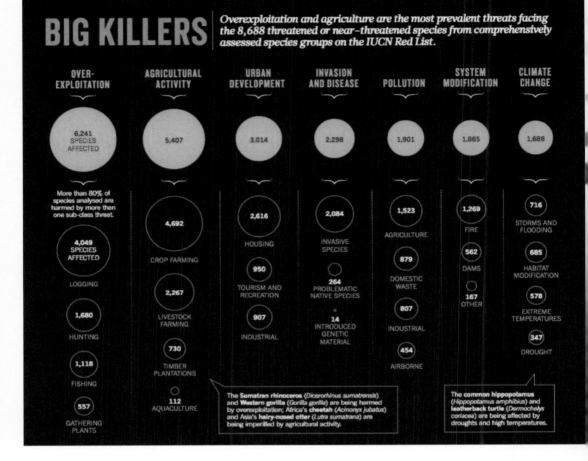

BIG KILLERS *Overexploitation and agriculture are the most prevalent threats facing the 8,688 threatened or near-threatened species from comprehensively assessed species groups on the IUCN Red List.*

OVER-EXPLOITATION	AGRICULTURAL ACTIVITY	URBAN DEVELOPMENT	INVASION AND DISEASE	POLLUTION	SYSTEM MODIFICATION	CLIMATE CHANGE
6,241 SPECIES AFFECTED	5,407	3,014	2,298	1,901	1,865	1,688

More than 80% of species analysed are harmed by more than one sub-class threat.

OVER-EXPLOITATION
- 4,049 SPECIES AFFECTED — LOGGING
- 1,680 — HUNTING
- 1,118 — FISHING
- 557 — GATHERING PLANTS

AGRICULTURAL ACTIVITY
- 4,692 — CROP FARMING
- 2,267 — LIVESTOCK FARMING
- 730 — TIMBER PLANTATIONS
- 112 — AQUACULTURE

URBAN DEVELOPMENT
- 2,616 — HOUSING
- 950 — TOURISM AND RECREATION
- 907 — INDUSTRIAL

INVASION AND DISEASE
- 2,084 — INVASIVE SPECIES
- 264 — PROBLEMATIC NATIVE SPECIES
- 14 — INTRODUCED GENETIC MATERIAL

POLLUTION
- 1,523 — AGRICULTURE
- 879 — DOMESTIC WASTE
- 807 — INDUSTRIAL
- 454 — AIRBORNE

SYSTEM MODIFICATION
- 1,269 — FIRE
- 562 — DAMS
- 167 — OTHER

CLIMATE CHANGE
- 716 — STORMS AND FLOODING
- 685 — HABITAT MODIFICATION
- 578 — EXTREME TEMPERATURES
- 347 — DROUGHT

The **Sumatran rhinoceros** (*Dicerorhinus sumatrensis*) and **Western gorilla** (*Gorilla gorilla*) are being harmed by overexploitation; Africa's **cheetah** (*Acinonyx jubatus*) and Asia's **hairy-nosed otter** (*Lutra sumatrana*) are being imperilled by agricultural activity.

The **common hippopotamus** (*Hippopotamus amphibius*) and **leatherback turtle** (*Dermochelys coriacea*) are being affected by droughts and high temperatures.

Sean Maxwell and others, *Biodiversity: the ravages of guns, nets and bulldozers,* **2016,** *Nature* **vol. 153–155.**

We are currently living through another extinction, this one much more rapid, and caused by humans. This graphic summarizes our understanding of which human activities are threatening the most species.

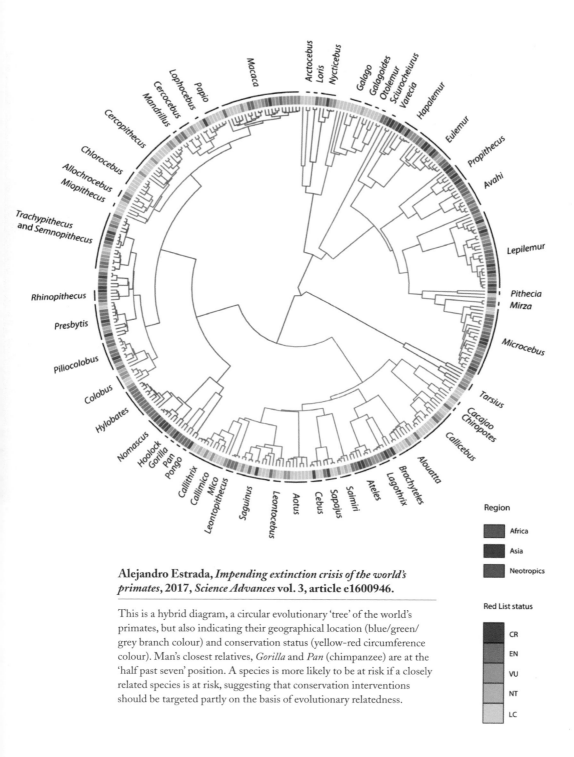

Region

■ Africa
■ Asia
■ Neotropics

Red List status

■ CR
■ EN
■ VU
■ NT
■ LC

Alejandro Estrada, *Impending extinction crisis of the world's primates,* **2017,** *Science Advances* **vol. 3, article e1600946.**

This is a hybrid diagram, a circular evolutionary 'tree' of the world's primates, but also indicating their geographical location (blue/green/grey branch colour) and conservation status (yellow-red circumference colour). Man's closest relatives, *Gorilla* and *Pan* (chimpanzee) are at the 'half past seven' position. A species is more likely to be at risk if a closely related species is at risk, suggesting that conservation interventions should be targeted partly on the basis of evolutionary relatedness.

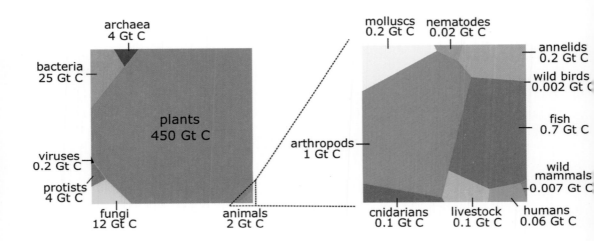

archaea
4 Gt C

bacteria
25 Gt C

plants
450 Gt C

viruses
0.2 Gt C

protists
4 Gt C

fungi
12 Gt C

animals
2 Gt C

arthropods
1 Gt C

molluscs
0.2 Gt C

nematodes
0.02 Gt C

annelids
0.2 Gt C

wild birds
0.002 Gt C

fish
0.7 Gt C

wild
mammals
0.007 Gt C

cnidarians
0.1 Gt C

livestock
0.1 Gt C

humans
0.06 Gt C

Yinon Bar-On and others, *The biomass distribution on Earth,* **2018,**
Proceedings of the National Academy of Sciences **vol. 115, pp. 6506–6511.**

One of the most fascinating zoological classifications was this recent assessment of
the *mass* of living matter in the whole world. The authors estimate there is a total of
550 billion tonnes (550 gigatons) of carbon locked up in living organisms, but that
only 2 gigatons of that resides in animals. Perhaps surprisingly, most of the Earth's
biomass is on land, mainly in the form of tree trunks – whereas most animal biomass
is in the sea. The right-hand block represents animal biomass only, approximately
half of which is arthropods (crustaceans, insects and spiders) and much of the rest
is fish. Humans weigh ten times more than all wild mammals combined and thirty
times more than all wild birds, but less than their own livestock.

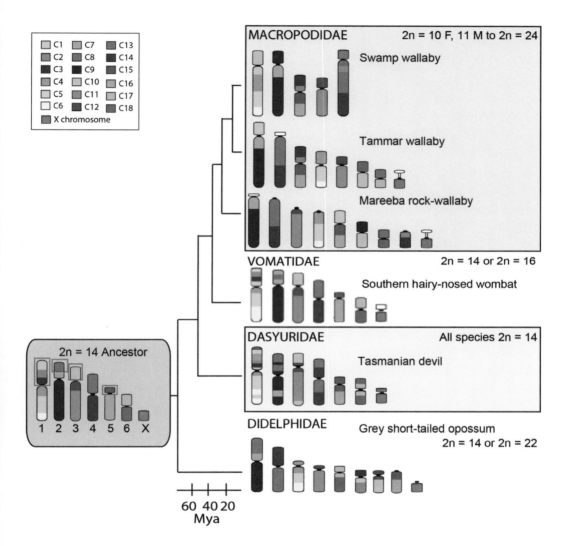

Janine Deakin, *Chromosome evolution in marsupials,* **2018,** *Genes* **vol. 9, p. 72.**

Not only do animals and their genes evolve, but the structure of their chromosomes changes over time, too. However, the reasons why chromosomes may split, fuse and exchange segments are poorly understood. Indeed, we do not know why species possess particular numbers of chromosomes nor why certain genes are carried on certain chromosomes. This study was an attempt to use marsupials' unusually large chromosomes to study their evolution. Some marsupial groups' chromosomes have changed little over time, while others have undergone dramatic rearrangement, yet the reasons for these differences remain unclear. (Mya means 'millions of years ago'.)

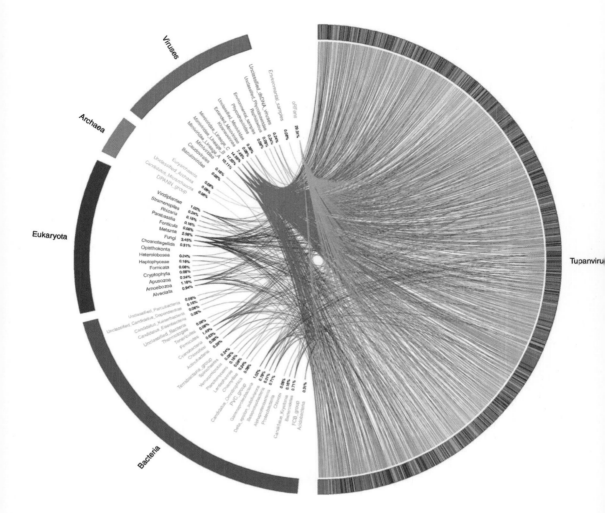

Jonatas Abraho, *Tailed giant Tupanvirus possesses the most complete translational apparatus of the known virosphere*, **2018,** *Nature Communications* **vol. 9, article 749.**

A major challenge of modern molecular biology is coping with the sheer volume of data it generates, so novel computer-generated visual representations are being invented to render these data comprehensible. This image is from an analysis of a virus with an unusually large inventory of its own genes – most viruses have few genes of their own and instead use the machinery of their host cell to replicate. The Tupanvirus's genes are mapped around the right-hand half of the circle, and are linked by arcing lines to the genes that most resemble them in the species on the left-hand side – other viruses (red); archaeans (pink); plants, fungi and animals (blue); and bacteria (green).

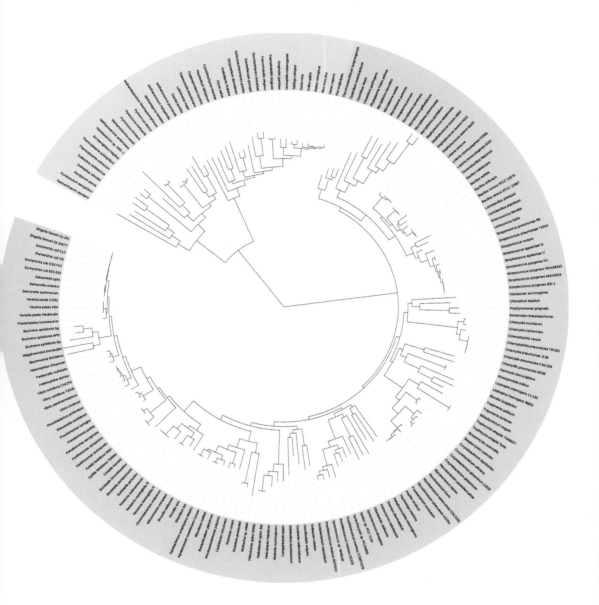

Interactive Tree of Life, 2019 (iTOL.embl.de).

Hosted by the European Molecular Biology Laboratory, the *Interactive Tree of Life* is not a single image, but an online tool to which scientists can upload their own taxonomic data and visualize it in a wide variety of formats, often visually appealing.

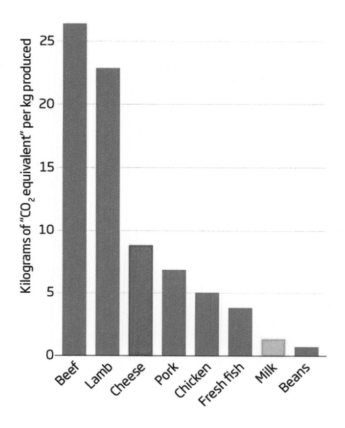

Graham Lawton, *The truth about cheese,* 2019, *New Scientist* vol. 3127.

It is now realized that keeping animals for food is one of the most important ways in which humans damage the environment. Adapted from the UN Food and Agriculture Organization's assessment *Tackling Climate Change through Livestock,* this graph compares the mass of carbon dioxide generated in the production of one kilogram of different meat and dairy products. Animal-derived food is always less energy-efficient than plant-based food, as it comes from higher up the food-pyramid (see page 204). In addition, the microbes in ruminants' stomachs produce methane, a more potent greenhouse gas than carbon dioxide.

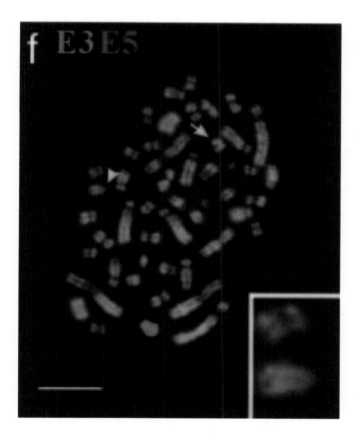

William Rens and others, *Resolution and evolution of the duck-billed platypus karyotype with an $X_1Y_1X_2Y_2X_3Y_3X_4Y_4X_5Y_5$ male sex chromosome constitution,* **2004,** *Proceedings of the National Academy of Sciences* **vol. 101, pp. 16257–16261.**

The platypus continues to challenge biologists' attempts to classify it. In most other mammals an individual's sex is determined by its pair of sex chromosomes – females having two X chromosomes and males having an X and a Y. However, platypuses have a complex chromosome complement, including what appear to be five pairs of sex chromosomes, not one, and their arrangement and inheritance are complex. These images are from a study using a technique called 'chromosome painting' to identify which segments these ten chromosomes share. Other studies suggest that platypuses lack the key gene that drives development of the male sex in other mammals.

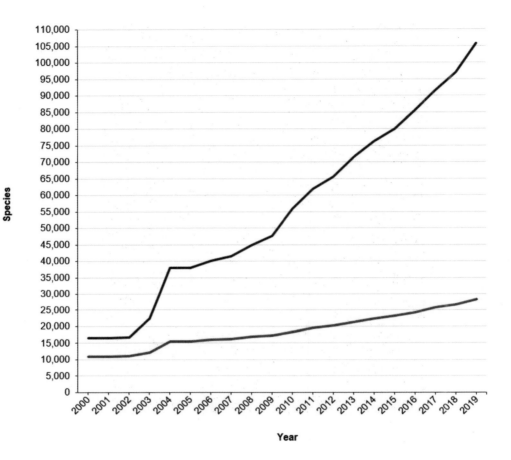

The IUCN regularly updates its list of the world's threatened species, but these data are often misinterpreted. The red line on the graph is the number of species assessed as threatened over the last two decades. Some of the increase is due to continued loss of habitat and other manmade causes, but much of it is due to a steady increase in the size of the IUCN's database (black line). The more species are assessed, the more species will be discovered to be threatened.

Total species assessed

Total threatened species

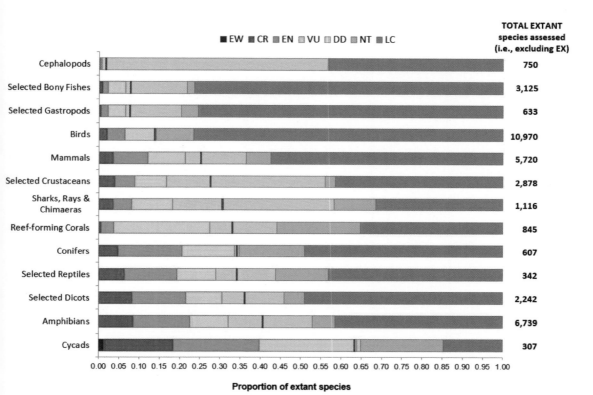

EW ■ CR ■ EN ■ VU ■ DD ■ NT ■ LC

Cephalopods	750
Selected Bony Fishes	3,125
Selected Gastropods	633
Birds	10,970
Mammals	5,720
Selected Crustaceans	2,878
Sharks, Rays & Chimaeras	1,116
Reef-forming Corals	845
Conifers	607
Selected Reptiles	342
Selected Dicots	2,242
Amphibians	6,739
Cycads	307

0.00 0.05 0.10 0.15 0.20 0.25 0.30 0.35 0.40 0.45 0.50 0.55 0.60 0.65 0.70 0.75 0.80 0.85 0.90 0.95 1.00

Proportion of extant species

International Union for Conservation of Nature, *IUCN Red List,* **2019; The
proportion of extant species in The IUCN Red List of Threatened Species.**

Human activities do not threaten all animal groups equally. In this graph, each horizontal
bar represents a taxonomic group of animals or plants, and is subdivided into colours
according to the level of threat its constituent species have been assessed to face. EW
(extinct in the wild); CR (critically endangered); EN (endangered); VU (vulnerable);
DD (data deficient); NT (near threatened); LC (least concern).

Index

Credits

Every attempt has been made to trace the copyright holders of works reproduced, and the publishers regret any unwitting oversights. Illustrations are generously provided courtesy of their owners, licensors, or the holding institutions as below:

Alamy Stock Photo/Everett Collection Historical: p.200

© 2019 American Association for Anatomy, by permission of John Wiley and Sons: p.189

© Amgueddfa Cymru – National Museum of Wales, photo Robin Maggs: p.109

Author's collection, and author's faculty collection: pp. 10, 44, 57 (bottom), 59, 78, 94, 95, 96, 97, 100, 101 (bottom), 103, 104, 114, 119, 123, 134, 135, 145, 146, 147, 149, 152, 156, 157, 162, 165, 168, 169, 170, 171, 174, 176, 178, 179, 180, 181, 182, 183, 186 (both), 187, 188, 190, 191, 193, 195, 196, 197, 198, 199 (both), 201, 202, 203, 208, 209, 210–211, 212, 213, 214, 215, 216, 218, 219, 220, 221, 222, 224, 226, 229

© 2014 Bachtrog et al. PLoS Biol 12(7): e1001899, under Open Access Creative Commons license (CC by 4.0) https://creativecommons.org/licenses/by/4.0/: p.237

Bayerische Staatsbibliothek München: pp. 32, 33 (Rar. 287, fols 4v, 5r), 38 (Res/2 Anat. 13, tab. III, p.69), 39 (Res/2 Anat. 13, p.67)

Biblioteca de Catalunya, Barcelona, under Public Domain Mark 1.0 license: p.28

Biblioteca Digital del Real Jardín Botánico, CSIC: https://bibdigital.rjb.csic.es: pp. 66–67

Bibliothèque nationale de France, Paris: pp. 24, 36, 37

Biodiversity Heritage Library (BHL): p.76

The British Library, London, under Public Domain Mark 1 (PDM 1.0) license: p.64

Carol Ballenger / David M. Hillis: p.233

CASG Slide #351069, Sara Mansfield © 2014 California Academy of Sciences: p.217

Chantilly, cliché CNRS-IRHT, © Bibliothèque et archives du musée Condé, Chantilly: pp. 27 (Ms 339, fo. 270), p.35 (Ms 139, fo. 33), p.34 (Ms 139, fos 3v, 30v, 29v)

The Dean & Chapter, Hereford Cathedral: Richard of Haldingham, *Mappa Mundi*, c.1300, details: pp. 17 (top), 26

D. Dunlop: p.85

David M. Hillis, Derrick Zwickl, and Robin Gutell, University of Texas: p.232

Evogeneao Tree of Life, 2008, 2017, © Leonard Eisenberg, courtesy of www.evogeneao.com: p.235

Dr Gabriel S. Bever, Center for Functional Anatomy & Evolution, Johns Hopkins University School of Medicine: p.240

Genes (Basel), 2018 Feb; 9(2):72, under Open Access Creative Commons license (CC by 4.0) https://creativecommons.org/licenses/by/4.0/: p.245

The Geological Society of London: pp. 92–93, 101 (top)

Getty Research Institute, Digital image courtesy of the Getty's Open Content Program: pp. 15, 29, 45

Getty Research Institute, Research Library (archive.org), via Biodiversity Heritage Library (BHL): pp. 54, 55

Graham Lawton, Guilty Pleasure: The Carbon Footprint of Cheese, *New Scientist*, 13/02/2019, © 2019 New Scientist Ltd. All rights reserved. Distributed by Tribune Content Agency: p.248

© 2011 Graphodatsky et al.; licensee BioMed Central Ltd, under Open Access licence (CC by 2.0): p.236

Harvard University Botany Libraries, via Biodiversity Heritage Library (BHL): pp. 52–53 (detail), 65, 84

Harvard University, Museum of Comparative Zoology, Ernst Mayr Library, via Biodiversity Heritage Library (BHL): pp. 58, 72, 73, 86, 108

Historic Maps Collection, Department of Rare Books and Special Collections, Princeton University Library: p.115

Howard Lynk – VictorianMicroscopeSlides.com: pp. 154, 155

iTOL.embl.de, Interactive tree of life, 2019, European Molecular Biology Laboratory: p.247

The IUCN Red List of Threatened Species, by permission: pp. 250, 251

The J. Paul Getty Museum, Los Angeles: pp. 22, 23

© Jeff Blossom, Center for Geographic Analysis, Harvard University: p.238

John Rylands Library, Copyright of The University of Manchester: p.25

Prof. Jonatas Abrahão, Institute of Biological Sciences, Universidade Federal de Minas Gerais, Brazil: p.246

Koninklijke Bibliotheek, The Hague: pp. 7, 30, 31

© Kurt Stueber (GNU Free Document License): p.161

Library of Congress, Washington D.C., licensed under CC0 1.0 Universal (CC by 1.0): pp. 133, 163, 164, 166; 136 (Geography and Map Division)

Maxwell SL, Fuller RA, Brooks TM, and Watson JE, 2016: p.242

MBLWHOI Library, via Biodiversity Heritage Library (BHL): pp. 107, 112, 144, 150–151, 153, 167, 204, 205

Mendel Museum, Masaryk University, Brno: p.142

Michigan State University Libraries: p.50

Missouri Botanical Garden, Peter H. Raven Library, via Biodiversity Heritage Library (BHL): pp. 74, 88

National Agricultural Library, Agricultural Research Service, U.S. Department of Agriculture: p.61

Natural History Museum Library, London, via Biodiversity Heritage Library (BHL): pp. 110–111, 130, 131

Nature Reviews, Microbiology, 2019, 17(4), 199–200 – fig. 1; courtesy of Prof. Ron Milo, Weizmann Institute of Science, Rehovot: p.244